Cross-Cultural Protection
of Nature and the Environment

Cross-Cultural Protection of Nature and the Environment

Edited by
Finn Arler & Ingeborg Svennevig

Odense University Press

Cross-cultural protection of nature and the environment

© The Authors and Odense University Press 1997
Printed by Narayana Press, Gylling, Denmark
ISBN 87-7838-347-1

The illustration on the cover of the book is:
Albrecht Dürer's watercolor *The Great Piece of Turf* (1503).
Reproduction with permission from
Graphische Sammlung, Albertina, Vienna

Published with support from
the Humanities Research Center 'Man and Nature'
and the National Forest and Nature Agency, Denmark

Odense University Press
Campusvej 55
DK-5230 Odense M

Tlf. +45 66 15 79 99
Fax. +45 66 15 81 25
E-mail: Press@forlag.ou.dk
www-location: http://www.ou.dk/press

Contents

Foreword 7
Ritt Bjerregaard: EU and environmental policy partnership
  – concerns and perceptions 11

Sustainability and the Good Life

Bryan Norton: A community-based approach to multi-generational
  environmental valuation 17
Avner de-Shalit: Sustainability and the liberal-communitarian debate..... 29
John O'Neill: The good life below the snow-line: pluralism, community
  and narrative 42

Biodiversity

Peder Agger & Peter Sandøe: The use of 'red lists' as an indicator
  of biodiversity 61
Kay Milton: Nature, culture and biodiversity 71
Merete Sørensen: Increase of biodiversity through biotechnology:
  genetic pollution or second order evolution? 84
Helle Tegner Anker: Biodiversity and the importance of the
  legal framework 93

Indigenous Peoples and the Protection of Nature

Paul Richards: Common knowledge and resource conservation,
  globally and locally 107
Darrell Posey: Utilizing Amazonian indigenous knowledge in the
  conservation of biodiversity: can Kayapó management strategies
  be equitably utilized and applied?..................... 119
Maj-Lis Follér: Protecting nature in Amazonia. Local knowledge
  as a counterpoint to globalization 134
Ingeborg Svennevig: Local peoples of the Western World
  – the introduction of local cultures in the Wadden Sea area......... 148

Cross-Cultural Conflicts and Cooperation

Poul Pedersen: Modernity, nature and ethics *161*

Finn Arler: Global partnership – a matter of friendship, reciprocity
    or mutual advantage? *176*

Tim Jensen: Religions and conservation. A survey. *192*

Randi Kaarhus: Policy discourses on environmental problems
    in Ecuador and Norway – a comparative perspective *206*

Klaus Lindegaard & Olman Segura: Trade offs in joint implementation
    strategies. The Central American forestry case *218*

Ulli Zeitler: Global solutions and local understanding.
    Conceptual and perceptual obstacles to global ethics and international
    environmental law. *232*

Contributors *245*

Foreword

States and people shall cooperate in good faith and in a spirit of partnership in the fulfilment of the principles embodied in this Declaration and in the further development of international law in the field of sustainable development.

Rio Declaration, Principle 27

The Stockholm Conference in 1972 marked the beginning of cross-cultural cooperation on environmental issues in modern times. Since then international cooperation on the cross-cultural protection of nature and the environment has grown considerably, and a large number of bi- and multilateral agreements, treaties and conventions in the field have been adopted, parallel to the establishment of national institutions dealing with these issues. Consequently, today environmental problems and the protection of unique, rare, threatened and/or vulnerable natural qualities are some of the most central issues on the international agenda.

During this process much learning has taken place, and resources from different cultural traditions have been brought forward, and attuned to each other as far as possible. One of the results is that several general principles have emerged, supplementing older more well-known principles of international law, like national self-determination and the principle of non-interference. Among the new principles one could mention the sustainability principle, the precautionary principle, the prevention principle, the *in situ* conservation principle, and the polluter pays principle. In the documents from the 1992 Rio Conference these principles are emphasized with much vigour, together with the need for a continuous effort to establish a global partnership around common problems and to adopt principles such as these to address them.

However, international declarations and conventions are not binding in the same way as national laws which are sanctioned by a state. Each nation will have to comply with the international agreements by making its own national regulations, based on its own interpretations of the adopted principles and their application to specific cases. Moreover, in most cases, the national implementation of the issues from the global agenda affects or involves local and indigenous people or peoples, who are located in areas of particular interest, and who, in some cases may consider their own way of life quite different from the mainstream national culture. So, even though the very adoption of global agreements is itself a victory for cross-cultural cooperation and unity, there is still considerable scope for cultural diversity and conflict within as well as between nations.

Such conflicts become more visible, and the attention to them grows as the implementation of local and national programmes increases. This again seems to go hand in hand with a more general and growing awareness of the need to take into account not only cultural differences, but also the principles of national and local self-determination. These are often presented as a right to voluntary and independent cultural and economic development. Since the first international declarations on environmental and nature protection issues were adopted, this awareness of the immanent dilemma between universal and culture-specific goals seems to have had a growing influence.

International declarations have to cover goals and ideals of very different nations and peoples, at the same time as the agreements necessarily have to rest on assumptions concerning a basic homogeneity of the people of the earth, and a common motivation to join in a global partnership for the protection of nature and the environment. How much attention should be paid to differences between national and international priorities, and how many special requests should be met, become something of a dilemma. There is also the ever present problem of how much homogeneity can be and should be assumed in relation to the interpretation of goals and ideals. After all, different people with dissimilar cultural backgrounds often tend to understand one and the same document very differently.

Within the overall problem of the international partnership on environmental protection, the articles in the present book focus primarily on three general themes. The first of these is the question of the general, and maybe culturally different aims of the protection of nature and the environment. When agreements are made, the final, carefully negotiated formulations are often vague and open to different interpretations. The meaning of commonly included phrases like 'protection of nature' or 'preservation of the environment' is anything but plain and unambiguous. The basic question is therefore, whether this protection is best understood as a matter of cleanliness, harmony, stability, diversity, integrity, or ecosystem health, whether or not human survival and welfare should be considered the core of the issue, and to what extent there are serious cultural differences concerning the interpretation. Obviously, this also concerns the issue of the extent to which humans are considered to have a unique moral status amongst other species.

The second theme involves the protection of nature and the environment from a temporal point of view and examines the issues of sustainability and responsibility with the future in view. Most international agreements nowadays speak of sustainability or sustainable development. These concepts can be interpreted in very different ways, however. One of the main questions is to what extent the term sustainability can be or should be understood and used in a way which is neutral towards different conceptions of the good, or whether this needs to be identified before the more precise obligations towards future generations can be formulated. Another, and more or less related, question is whether or not the request for sustainability is to be understood primarily in terms of one's particular multi-generational community and its special values and conceptions of the good. This again is related to the question to what extent the issue of sustainability can or should be linked to global obligations proper, oriented towards future people as such, i.e., towards people with an indeterminate relation to our own conception of the good.

The third theme is concerned with the implementation of global aims of environmental protection across cultural differences. The problem is how to combine the goals and perspectives of those locally affected, with the demands and requirements of the global partnership. What may be seen in some quarters as a question of implementing a general goal efficiently in a specific setting, may elsewhere frequently be considered as an unacceptable, and often quite incomprehensible intrusion into the local culture. If, on the other hand, the global aim is to be realized, in many cases the local people have to be involved on the same side as the conservationists. In some instances, however, the national government is not working on that side. There is considerable potential for conflict and disagreement but the many examples of communities working in unison and reconciling their differences, within both local and global frameworks, should not be overlooked.

The book opens with a short article by the EU-Commissioner Ritt Bjerregaard on problems related to cultural differences within the European Union. The remaining articles are grouped under four section headings. The first section has *Sustainability and the Good Life* as its general theme. Bryan Norton, John O'Neill and Avner de-Shalit, all of them with a background in philosophy, discusses some of the main differences between various approaches to the concept of sustainability, especially welfare economic, liberalist and community based approaches, and their relation to the variety of understandings of the 'good life'. A common theme is whether the protection of nature and the environment should be based on thin or thick understandings of the good. The three authors all agree that a thin and seemingly neutral understanding of sustainability is inadequate.

The contributions in the second section all have *Biodiversity* as their focal point. Biodiversity has become a key concept in environmental protection over the past decade. It is not a simple concept, however, and it is anything but obvious what should be preserved, why and how. All the contributions in this section make it clear that the preservation of biodiversity is more than just a question of counting and keeping the natural number of species. They also bring attention to the practical difficulties which emerge when the general aim of preserving biodiversity is to be specified and implemented in a local setting. The contributors in this section, Peder Agger and Peter Sandøe, Kay Milton, Merete Sørensen and Helle Tegner Anker represent a wide spectrum of disciplines: biology, philosophy, law, and anthropology.

The section on *Indigenous Peoples and the Protection of Nature* discusses the relationship between global agendas and local and indigenous knowledge, experiences, aims and understandings related to the natural environment in specific areas. This section opens with a discussion by Paul Richards on the common ground between natural science and local knowledge. Conflicts which arise out of differing local and global perceptions of the issues involved and how they are actually resolved – in areas as different as the Amazon basin and the Wadden Sea – are explored by Darrel Posey, Maj-Lis Follér, and Ingeborg Svannevig, all of whom have a background in either anthropology or ethnobotany.

Finally, the section on *Cultural Discrepancies and Cooperation* investigates various aspects of the present attempts to establish international and global partnerships to protect nature and the environment. This section explores a wide range of issues: the development of the global partnership; views on what nature is and how it can best be explained, according to differing contemporary and pre-modern perceptions as to what precisely it is; similarities between the world religions, and, finally, what posssibilities exist for the joint implementation of global strategies across national and cultural borders, involving different partners. Poul Pedersen, Finn Arler, Tim Jensen, Klaus Lindegaard and Olman Segura, Randi Kaarhus, and Ulli Zeitler thus contribute with their analyses, which are related to anthropological, philosophical, religious, historical and economic dimensions of the overall theme.

Earlier versions of the articles included in this volume were presented as conference papers in May 1997 at the "Cross-Cultural Protection of Nature and the Environment" seminar, hosted by the Humanities Research Center, at Odense University, Denmark. The intention behind the seminar was to create a forum where various kinds of researchers, administrators and politicians could debate issues across national and disciplinary boundaries in an attempt to bridge the gap between common global and exclusively local perceptions of the issues involved. The final result, which can be found in this volume, is a combination of philosophical discussions of fundamentals and anthropological analyses based on a sense of the tangible, complemented by contributions from various historical, natural and social sciences. Thus, a kind of cross-cultural and cross-disciplinary partnership between different approaches to the protection of nature and the environment has been established.

EU and environmental policy partnership – concerns and perceptions

Ritt Bjerregaard
Original text condensed by the editors

I think that the issue you have chosen to explore is a very important one. It is all too easy for those of us involved in policy-making to focus on what we think are the big environmental problems in our own cities, regions and countries. Nobody needs to be reminded that finding the best solutions to environmental problems usually requires cooperation across borders. What is often forgotten is that in order for these solutions to be found, they must make sense to everyone involved.

It is difficult enough to get agreement on environmental measures in a situation where everyone comes from a similar cultural background. How can we reach agreement between people and nations who may have very little in common? Firstly, the needs of people living in different areas vary. European urban citizens may need a place they can have a relaxing picnic on a Sunday, while a farmer in a developing country may need to farm his land to avoid the possibility of starvation. Secondly, environmental surroundings are simply different depending on where people live. Some are used to living in areas with a scarcity of water, some live in densely populated areas, others live in harsh, isolated climates. These different needs and situations are often translated into different concerns. The combination of these differences leads to a certain view of the environment and problems related to it which vary from place to place. Thus we can speak about cross-cultural perceptions of the environment.

We need to pay attention to these differences to find environmental solutions which are more complete and therefore, ultimately more successful. Dealing with people of different perspectives forces us to look at different facets of an environmental issue.

Partnerships

Partnership is one of the most exciting concepts in the environmental policy field and although it sounds basic, I think it is only beginning to be fully explored. You may have heard the term subsidiarity being used in Euro-circles – it strikes me as a fancy way to describe a thoughtful partnership. Actually, it may interest you to know that the principle of subsidiarity was first applied in the European Union to environmental issues. The basic idea of subsidiarity is that the Union should only act when it is necessary and when Member States, acting on their own, cannot address a certain problem. At a practical level, this means that many levels of government need to be involved in finding solutions to environmental issues. It is a recognition that each level of government – local, regional, national, Community – holds a piece of a puzzle.

This is recognition on an institutional level that no level of government can provide the complete answer to. I think that a good example of this is air quality. Improving air quality is a high-profile environmental issue in many cities. From a single market point of view, there are measures which have been taken at Community level because of their cross-border implications. The Community has therefore adopted what is known as the Auto-Oil programme, which will ensure consistence in emission standards and in the quality of fuels used throughout the community. We will reduce emissions and get cleaner air to breathe. Any supplementary measures are left to other levels of government. Thus, Athens, The Hague, Helsinki, Odense and other European cities are developing programmes depending on the needs of their city to complement what is happening at Community level, from creating pedestrian areas, to bicycle networks, to improving local public transport, to building more centrally-located housing.

Another good example of the use of partnership is acidification. Many Member State governments have taken measures to deal with the effects of certain chemicals on their environment. However, part of the acidification problem comes from areas outside the control of a national government since of course air circulates quite freely across borders! This is particularly the case for Scandinavia where much of the acidification comes from emissions in countries to the South and West. Action was therefore also necessary at International and Community level. A significant part of this has been a Directive on Large Combustion Plants which imposed limits on the amount of NO_2 and CO_2 which could be released into the atmosphere. This is a good example of where Community Member States have accepted the responsibility, and cost, of resolving major environment problems which occur outside their own territory.

The EU and the world

Although the EU must deal with cross-cultural issues, it is obvious that the differences between countries increase greatly beyond the borders of the Community. This makes it very difficult to reach accepted standards and, I think, all the more crucial to focus on the issue of environmental perceptions.

Of course, there are international success stories. One of the greatest is the Montreal protocol on substances depleting the ozone layer. Another noteworthy example is the Basel convention on the transborder shipment of waste, where developed countries have accepted responsibility for the management of their own hazardous waste rather than to take the easy route of shipping it to third countries. In these cases, there was agreement among most of the states causing the greatest damage that something had to be done. In other cases, however, different understandings between nations on environmental issues can slow progress or even lead to conflict.

Despite great efforts on the part of several countries over the past years, many of the agreements reached in Rio remain empty. For instance, although many countries share a concern over deforestation, there has been little concrete international action to combat the issue in a way which can meet the needs of the countries in which tropical forests are found. A forest convention would be a good way to address this issue, but the hesitation of many countries shows just how much work we have to do. Biodiversity is also a serious environmental concern. Again, this is an issue which will require intense international cooperation if we are to make the progress we want. Yet it is also an issue which meets head-on the issue of different perceptions of the environment. Biodiversity is not often of much concern to people who are seeking to improve very minimal standard of living and their governments are often reluctant to make it a priority issue.

The European Community, as you may know, has been involved in the G-7 Pilot Programme in Brazil. In fact the Community is the second largest donor to the programme after Germany. While I think the pilot programme has been successful at raising awareness, the funding for this programme is not very much compared to the problems and challenges posed in trying to introduce sustainable development in Amazonia.

Conclusion

I would like to conclude with my belief that we must always find environmental solutions which can address the underlying causes of differing environmental perspectives. Only in this way will we be able to expand the scope of international common action on the environment. In some cases, this means creating flexible agreements which make sense to all and are implemented according to national custom. In other cases, this means addressing the issue of needs, be they economic or social. This is the most difficult question which faces us: as an international community, how far are we willing to go in order to accommodate diversity?

I know that in order to count as many successes at international level as we have had at Community level, we will have to become increasingly conscious of cultural differences in needs, concerns and perceptions. This is where your work as researchers is crucial to us as policy-makers attempting to make progress on cross-border environmental issues. I encourage you to continue your efforts and I wish you all good luck with this seminar.

Sustainability
and the Good Life

A community-based approach to multi-generational environmental valuation

Bryan Norton

Introduction

How should we conceptualize and evaluate environmental policies that have multi-generational as well as short-term impacts? This question lies at the heart of attempts to define and measure sustainability. Unfortunately, in my view, most discussions of sustainability fail to ask this question on the profound level it requires, usually because the participants in the discussion all assume that sustainability will be measured using a broadly utilitarian comparison of welfare opportunities across generations; more specifically the methods of welfare economics are usually used to operationalize those comparisons. I will argue that, while economic valuation may capture an important aspect of environmental values, its methods are ill suited, by their very logic, to address questions of intergenerational equity. What is needed, if we are to have a reasonable discussion of the benefits of living sustainably, is a way to compare values across generations.

In this paper, I explore an alternative to economic measures as indices of cross-generational sustainability. Eventually, I will propose an index based on the idea that what we owe the future is to refrain from narrowing their range of choices in their struggle to survive, from narrowing the opportunities they have to experience a rich and meaningful life, and from destroying the natural conditions necessary to continue a community's valued style of life. But before taking this bold step, the way must be prepared by stating the logical characteristics and formative assumptions of the economic model, and by showing how those assumptions could be altered to make the model more sensitive to important, intergenerational considerations that will be overlooked if we understand intergenerational equity as a simple matter of income comparisons. I will argue, in opposition to the economic, welfare paradigm, that we may owe the future some quite specific protections and that the value of these protections can only be captured within a multi-generational, community-based model of environmental protection, a model in which each generation respects the right of the next to face a range of options and opportunities equal to those in earlier generations. This approach to valuation is based on the intuition that a freedom-loving community will place a high value on the perpetuation of the conditions necessary for future members of the community to enjoy an equal or greater range of choices as they negotiate their own survival through creative adaptation.

Freedom is about choices; choices require options. So options are the prerequisite of true freedom; if I control the options you chose among, I control the range – and the substance – of your free choice. Every human community represents a unique negotiation between a people and its natural setting. If I destroy the physical context and the resource base that is constitutive of a community's sense of who they are, then I will have destroyed a community – or at least I will have created a discontinuity between the past and present, and have strained the fabric of intergenerational community. The purpose of this paper, then, is to explore whether this intimate conceptual connection between the range of options available and an unquestioned value – freedom of choice – might support a new approach to long-term environmental valuation and management, an approach that is more sensitive to community-based, multigenerational concerns.

Economic sustainability, weak and strong

The interdisciplinary debate now raging over the best way to define and measure sustainability of policies and practices centers on whether 'weak' sustainability, as defined within the theoretical structure of growth theory as it is embedded in welfare economics, is adequate to express all of the obligations that are owed the future with regard to protection of resources. According to the weak sustainability theory, as explained by Robert Solow, earlier generations owe no particular resources to the future – as long as each generation maintains a non-declining stock of capital (defined generally as natural resources, technological abilities, industrial capabilities, and wealth), it will have fulfilled its obligations to the next.[1] Weak sustainability, then, merely requires that each generation adopts a 'fair savings rate,' which requires that savings are adequate to provide a nondeclining stock of capital for investment and efficient production for future generations. According to this view, human-built capital, such as technology and infrastructure, can be an adequate substitute for resources that are used up or degraded. By treating human-built capital and natural resources as 'fungible', economic growth theorists thereby propose an unstructured bequest package as an adequate response to future needs. As long as the future is as well off as we are, measured in terms of accumulated, and undifferentiated, capital, they have no basis for complaint.

Ecological economists argue, in opposition, that weak sustainability is inadequate because it incorrectly treats natural and human-built capital as substitutes, when they are, in fact, complements. They argue that depletion of natural resources, even if replaced with equal amounts of human-built capital, can increase prices and, by driving up prices, reduce the real income and opportunities to enjoy welfare of future generations. Indeed, they claim that this process is occurring even now; they therefore insist that we maintain accounts that measure 'natural capital' and accept an obligation to protect adequate amounts of natural capital, as well as adequate amounts of general, undifferentiated capital. They therefore advocate what we can call a 'structured' bequest package.

While this spirited debate has led to increased insight, my goal here is to emphasize the ultimate similarity of weak and strong sustainability in that both systems of analysis use a single, economic scale of value. Within this shared scale of value it is in principle possible to *compensate* the future for any losses to natural capital. Their analysis therefore retains the main features of marginalism and compensability so familiar in welfare economics. They never question, therefore, that losses to the future can be represented in terms of present values, and they expect to represent losses of natural capital within a synoptic computation of aggregated individual welfare values.

To see this similarity, consider the 'green' accounting system proposed by Daly and Cobb.[2] Daly and Cobb note that national accounts, even when measured inadequately as Gross National Product (GNP), are usefully corrected for depreciation of capital and that, at least over time, consumption must be limited by the Net National Product for that year. Daly and Cobb then argued (along with other corrections that need not concern us here) that it makes equal sense to correct national accounts for 'depreciation of natural capital'. So, by analogy to the 'adjustment account' kept for depreciation of capital, Daly and Cobb suggest another adjustment account to measure loss of natural capital as losses to be deducted from general accounts of all wealth including technological development, factories, etc. This deduction from national accounts, they conclude, is justified *"simply to gain a better approximation to the central and well-established meaning of income."*[3] To follow up on this conclusion, they operationalize this notion in their 'Index of Sustainable Economic Welfare,' which is presented as an extensive Appendix to the book. In their argument, they dispute the conclusion of economists Nordhaus and Tobin, on the point of substitutability, concluding that resource depletion is already affecting income, and that *"since 1972, the stagnation of productivity for about a decade are signs of the effect of rising real resource costs, particularly energy resources."*[4] They say, *"We have thus deducted an estimate of the amount that we would need to set aside in a perpetual income stream to compensate future generations for the loss of services from non-renewable energy resources (as well as other exhaustible mineral resources). In addition, we have deducted for the loss of resources such as wetlands and croplands ... This may be thought of as an accounting device for the depreciation of 'natural capital ..."*[5] The value of resources for the future is equated to an 'income stream' from a trust fund that will compensate their loss of resource availability.

At this point in the argument, however, the differences between strong and weak economic sustainability become less and less tangible, even as their similarities – especially the willingness to compensate the future at the present exchange rate for a lost future opportunity – become more evident.

I intend to argue, especially in the following parts of this paper, that one cannot effectively represent, or measure, losses to the future using the paradigm of welfare economics and that we must specify particular features of the natural world – elements or processes – in non-economic terms that we owe it to the future to protect. In order to set the stage for this argument, I briefly summarize five characteristics of the economic conception of value which make it ill suited to capture the full-blooded concept of sustainability that is embraced by most environmentalists – or at least by those who have not bought into the economic approach to environmental valuation and measuring sustainability.

1. All economic values are expressed as present values. One of the unquestioned strengths of economic analysis is that, given certain assumptions, it is possible to aggregate and compare quite disparate values, including values that are experienced at different times. In particular, economic analysis, through the application of discounting and corrections for inflation, can apparently compare the present values of future experiences. Thus it is possible, for example, to express the value of a future option as the amount one should be willing to pay in the present to hold open that option. While this approach is often useful if we wish to examine individual behaviour, it is much more problematic, ethically, if one applies this reasoning to decisions affecting multiple generations. The value of an option to future persons, on this application, would be the willingness of present persons to pay to hold that option open for the benefit of future persons. Values that will be experienced in the future must therefore be expressed as the present market value of an option that may not be exercised until far in the future. Thus, all generational saving must be considered by economists as a type of altruism since, by paying to protect an option in the present, one transfers a benefit from oneself to a person not yet born.

2. Economic values are individualistic values. Perhaps the most central of all

economic assumptions is the assumption that the values of 'rational, economic man' are the values of self-interested individuals. Apparently, then, this feature of economic valuation ensures – in conjunction with 1., above, that 'obligations' to the future are not really 'obligations' at all: they are entirely optional. An obligation to live sustainably is only as strong as the 'preference' of present individuals to act altruistically with regard to the future. Further, commitment to individualism entails that the economic model of valuation countenances no obligation to a 'community' or 'common good', except insofar as these concepts can be reduced to an aggregation of individual goods.

3. Economists assume 'consumer sovereignty'; values are fixed, static. For methodological reasons, economists (in order to ensure that individual choices in markets can represent welfare without confounding complications) insist that each individual is the best judge of his or her own preferences[6] and, accordingly, economists take preferences (as expressed in market behaviour or in responses to hypothetical market situations) as 'given' and unchanging for the period of analysis. But it has been pointed out that, however useful this simplification is when one is measuring and analyzing values as expressed in short periods of time, it is obvious that human preferences do change over time, especially if the time period involved is considerable.[7] Economic valuation studies therefore provide at best a static 'snapshot' of a dynamic process of preference formation, reconsideration, and reformation. While this dynamic aspect might fruitfully be ignored when one examines short frames of time, it causes significant distortions when economic valuation is applied to longer, intergenerational contexts.

4. Economic values are place-neutral. In order to ensure that measurements of values are agreeable, economists treat the place at which a change in value occurs as irrelevant to its importance. For example, when computing the GDP for a state or country undergoing rapid economic change, jobs and opportunities lost in one area or region will be 'compensated' if jobs and opportunities are created elsewhere. Thus, it is stated that the US. States of Oregon and Washington are enjoying 'economic growth' and 'rapid development', even as whole cities and towns in out-state regions, dependent on logging and exploitation of other natural resources, are being closed down. The devastation of whole towns is offset by the rapid development of high-tech industries in Portland and Seattle. As economic gains and losses are aggregated over larger and larger areas, the spatial location of economic opportunities becomes irrelevant; economic analyses therefore take no account of the importance citizens place on commitments to their 'home place'.

5. Economic losses are always compensable. As was noted in the above discussion of Daly and Cobb's "Index", even if one identifies some aspect of the

environment as natural capital, the future can be compensated for its destruction by setting up a trust fund, the income from which will compensate for lost income/higher prices affecting future persons. Thus, while it seems that identification of natural capital places greater constraints on present action than are implied by weak sustainability theory, the commitment of economists to the view that all goods have a price undermines these apparent constraints. All value, on the economic view, can be expressed in present dollars; therefore, losses which occur in the future due to erosion of natural capital in the present have a present-dollar equivalent, ensuring that all future losses are compensable by trust fund or other monetary device.

This complex of features of value in the economic paradigm apparently ensures that any form of sustainability that is conceived and measured only in terms of economic value will ultimately allow compensability as a substitute for actually protecting resources, even resources that are designated as natural capital. It is therefore reasonable to place both strong and weak sustainability theory within a larger classification of 'economic sustainability' to emphasize that they share a commitment to an economic approach to valuation.

Why economic sustainability is not enough

Economic sustainability, the idea that we ought not to impoverish the future by over-consuming and leaving the future with lesser opportunities to enjoy welfare levels comparable to ours, is no doubt a *necessary* condition that must be fulfilled if we are to be fair to the future. The question, then, is whether fulfilling this obligation to the future is sufficient to capture the full range of concerns that motivates environmentalists' to advocate sustainable lifestyles. In this part, I explore the implications of this question by varying the characteristics listed in the first section.

1. Human values unfold at different scales of time. In contrast to the economic idea that all values can be expressed as present values, I believe it is more accurate to attribute to human beings an ability and inclination to value things on multiple, irreducible temporal scales. The best example to illustrate this point is a constitutional convention. If the people of a country, for whatever reason, determine that it is time to write a new constitution to govern their political life, the citizens will go to great pains to choose representatives who will take a long view of things, who will distance themselves during deliberations from their own, immediate economic interests, and who will think seriously about constructing political institutions that will serve the country and its citizens well over many generations. In so choosing, the citizens are unlikely to

ask, "How much are we, the present citizens, taken collectively, willing to pay in the present to ensure the rule of law for future generations?" Some values, such as the consumptive values of economics, are calculated with quite short 'horizons', while other values, especially the values of institutions that constitute political life in the community are understood in a much expanded, multi-generational frame of time. Concerns about sustainability, I believe, should be thought of analogously to such constitutional values. Whereas a constitution designs political institutions that should last for generations, and determine what type of community will evolve over time, concerns about sustainability address an analogous question of how a community will 'constitute' itself within the larger system of nature.

2. Sustainability values are community-oriented, not individually oriented. As is clear in the example of the constitutional convention, concerns about the long-distant future – both political and resource availability concerns usually associated with sustainability – are best understood as community-based values. Mark Sagoff, and others, have therefore argued that economic valuation fails to recognize a crucial distinction between individual decisions made 'as consumers' – decisions that accept basic market conditions – and those made 'as citizens'.[8] It is a mistake to assume, as economists do, that all values associated with sustainability should be conceived and measured as self-regarding decisions of individuals – sustainability is a community-oriented value. We must recognize also that, in many contexts, people acting as citizens choose policies they believe to be good for their community, rather than for them, personally. The choice to act sustainably is a community-oriented, not an individually-oriented decision.

3. Values – both individual preferences and community-oriented social values change across time. One function of environmental and other social leaders is to encourage consideration of the environmental impacts of our consumptive and other values, and to encourage public discussion and debate regarding environmental values and goals. Not all values are equally environmentally friendly and we are as justified in trying to re-form those values as we are in trying to change the values of racists. In order to develop sustainable communities, it may be necessary to engage in discussion and education programs aimed at encouraging citizens to reform their individual preferences in favour of longer-term, community-oriented values.

4. Environmental values are place-based and sustainability must be sensitive to locality and place.[9] When people talk about sustainability, I think they are talking about sustainable communities which, by sensibly managing their natural and human resources, survive and thrive in a particular and distinctive place for many generations. This form of sustainability is as much about build-

ing fair and responsive institutions as it is about ecologically sensitive management; the point is that real people experience environmental problems in a specific 'place', and they evaluate changes in a similarly perspectival manner. Since plants and animals – the ecological communities that form the context of human communities – are adapted to their local place, it makes sense to expect that the natural values expressed by members of a community will likewise be place-dependent and expressive of local character.

5. Some destruction of resources results in noncompensable losses imposed on future generations. On this approach, we could ask: Are there possible damages we could inflict on resources that would make people of the future worse off than they would have been if those resources had been protected, *even if individuals in the future are richer than we are*. This definition captures a stronger sense of sustainability by which persons in the future can be judged worse off because certain options have become unavailable to them. For example, if natural areas are all lost and replaced with Disney-style 'virtual' natural areas, a person in the future might be considered worse off even if they have adequate income to visit, or even own, virtual natural areas.

Dynamic, multi-scalar, community-based evaluation

I begin my exploration of an alternative approach to evaluation by quoting the sincere and heartfelt expression, by a locally active environmentalist from Southern Tennessee, of what was to me a convincing 'environmental value'. The activist was expressing his frustration at a series of governmental and private decisions, decisions that seemed to make it more and more inevitable that large multi-national corporations would be allowed, even encouraged, to construct mega-mills along the Tennessee River, huge mills for grinding hardwood forests into chips for export, mostly for the Asian market. The extraordinary scale of these mills would ensure that virtually all of the remaining hardwood forests in the Southern Appalachians will be 'chipped out'. The activist said, "*If they let the chipmills in, they'll scour the Southeast, and replant fast-growing pine in straight rows. I grew up in a hardwood forest. We like our hardwoods. I'll fight to stop them, but it seems pretty hopeless, with the government talking 'jobs', and the big Japanese money behind the mills.*" This, I believe, was a sincere and heartfelt expression of an environmental value. It is a value that Solow cannot capture with his aggregations, and the activist did not even bother to ask whether he or his children might increase their income if the mills are built. The value he was expressing cannot be measured in present values, and it has little to do with income and opportunities to consume more. What exactly is this value?

One might characterize such values in a number of ways – aesthetically, perhaps – but I propose that we characterize this value as a 'place-based' value;[10] place-based values emerge from a local dialectic of culture and nature; some of these place-based values are 'constitutive' – they express the distinctive identity of the place. To allow them, or the context that gives them meaning, to be destroyed is to our activist to accept cultural death at the hands of multi-national corporations. The sum total of the identity-constituting values can perhaps be summed up as 'the integrity of a place'. Some actions threaten the integrity of a place; others do not. A place, while not an 'organism' in any technical sense, emerges organically as a group of people struggle to attain or retain a cultural identity, a cultural identity that is appropriate, adaptable, fitting, and also aesthetically pleasing *as a way of inhabiting their own chosen geographic place*. The notion of place I adopt, then, is thoroughly cultural, and cannot be defined simply by understanding the natural system independent of human-nature interactions.

I believe that the values expressed by the activist from Tennessee represents a commitment to his past and a commitment to the future of the community – both human and ecological – he loves. He and his parents, and theirs yet earlier, had 'adapted' to hardwoods. The hardwood forest constitutes, even as it contextualizes and gives meaning to, the value as expressed. To this activist, living in an area forested with hardwoods was a 'constitutive' value; it makes him who he is, and represents also his aspirations for his children. Somehow, he can't see *his* children hunting, hiking, living the cultural life that has evolved within his community in a pine-plantation context.

I propose that we interpret this feared loss as a desire that subsequent generations face, in their struggle to survive and flourish, opportunities equal, and also similar, to those we have faced. The next step toward a more nuanced and scale-sensitive approach to evaluating projects and policies would be to express this desire as a value placed on certain opportunities being open in the next generation. Likewise, a negative value would be placed on choices that eliminate important future opportunities. We have therefore established a connection between impacts of today's environmental decisions and the range of choices available in the future. The next step, only vaguely described here, might be to choose one or several physically describable indicators that track the performance of a culture in sustaining resource-based activities and in finding an ecologically harmonious place within a landscape over multi-generational time, and to develop a model for comparing consequences of decisions *both* for economic well-being *and* for impacts on the range of options to be faced by members of that culture in the future.

One consequence of the shift from a narrowly economic to a more complex and multi-faceted conception of environmental values is a shift to a system of evaluation that is more pluralistic and capable of measuring a much wider range of trends. The point is not to give up economic analysis of environmental problems – we noted, above, that weak sustainability is a necessary condition of fulfilling our obligation to the next generation. But now economic analyses are included within a battery of indicators, indicators which might be applied simultaneously to a single problem, or deployed differentially, based on objectively describable conditions that might 'trigger' the application of a particular indicator that is of particular interest in evaluating particular policies in particular situations.[11] The point is that we now enter the discussion of how to achieve sustainability with a number of measures and indicators, and ideal policies must do well on all of the applicable criteria. What we look for in policies, then, is not that they score the highest on one, reductionistic measure such as economic efficiency, but rather that they are 'robust' in doing reasonably well on a range of criteria.

Another consequence of accepting this pluralism and complexity of evaluation is that it is no longer necessary to express evaluations as assignments of worth to 'environmental commodities'. Ecological description of nature tends to concentrate on processes and energy flows and a more ecologically sensitive evaluative tool might evaluate varied 'development paths'. For example, one might produce several computer simulations that would show the expected changes in the landscape that would result from various development paths, by running the clock forward, showing citizens what the countryside will look like in the future if chip mills are encouraged, and what would happen if they are banned and other forms of development, such as tourism, or high-tech industry are encouraged. If this were possible, then citizens could align and compare expected economic outcomes of various paths of development with consequences to their life-style, and make more informed decisions.

A third consequence of adopting this approach is that it should be possible, once we have progressed to develop more sophisticated measures of several types, to point out to people that the preferences they express and behave upon in their day-to-day life are incompatible with protecting longer-term and more community-oriented values. For example, viewing a simulation of development based on automobile-based transportation system and the impacts of such upon the landscape as expressways proliferate, might convince citizens that their preference for automobiles over public transportation or a bicycle must change if they are to leave for their children a beautiful and comfortable landscape.

Finally, while the proposed approach includes a concern for economic well-being and development, it also focuses on the longer-term question of maintaining valued opportunities that allow people in the future to choose among valued lifestyles. Especially, it seems important that certain options and opportunities – ones that are essential to maintaining community identity across time – are maintained so that the values of cross-generational continuity of communities can be maintained.

Conclusion

I have listed some reasons to think that the economic model of valuation, and the economic notion of sustainability that goes with it, provide at most a necessary, or minimal, condition on fair treatment of the future; but this economic approach is ill suited to capture the full range of values that most environmentalists would include in their concerns to build 'sustainable' communities. In contrast to the reductionistic approach of welfare economics, I have suggested a multi-criteria approach that rates policies according to more than one criterion, and a good policy will be one that is robust with respect to short and long-term criteria. Once we adopt a multiscalar and pluralistic value system, one type of non-economic value is the value we place on individuals having a range of opportunities among which to choose, both in their economic life and in their aesthetic and personal life. One can harm the future either by making them so poor they cannot enjoy the opportunities available at the time that they live, or we can harm them by narrowing the range of options they have for inspiring and enlightening experiences.[12] So, the protection of landscape-level features, such as maintaining a significant amount of hardwood forest cover in a region, may hold open important options and protect the freedom and opportunity of the next generation. Finally, I have also suggested that, if a community – conscious of its natural history and comfortable with the practices and adaptations they have developed as a cultural community – has carved a viable and satisfying existence within a local, ecological context, they may feel an obligation to protect crucial ecological features that support and give meaning to their culture. Sustainability, in this full-blown sense, includes – in the face of possible cultural annihilation – an obligation to protect the natural, ecological conditions necessary to maintain a locally valued lifestyle, and to maintain the conditions of shared experience across generations within a cultural community.

Notes

1. Solow, Robert: "Sustainability: An Economist's Perspective", in R. Dorfman and N. Dorfman (eds.): *Economics of the Environment: Selected Readings*, New York: W.W. Norton and Company, 1993, pp. 179-187.
2. Daly, H. and J. Cobb: *For the Common Good*, Boston: Beacon Press, 1989. See, especially, the Appendix.
3. *Ibid.* p.71.
4. *Ibid.* p.410.
5. *Ibid.* p.411.
6. Norton, B.G.: "Economists' Preferences and the Preferences of Economists", *Environmental Values* (3)1994: 311-332.
7. Norton, Bryan, Robert Constanza and Richard Bishop: "The Evolution of Preferences", *Ecological Economics*, forthcoming
8. Sagoff, Mark: *The Economy of the Earth*, Cambridge and New York: Cambridge University Press, 1988.
9. See Bryan Norton and Bruce Hannon: "Environmental Values: A Place-Based Theory", *Environmental Ethics* 19 (1997), pp. 227-245.
10. *Ibid.*
11. Norton, B.G.: "Evaluating Ecosystem States: Two Competing Paradigms", *Ecological Economics* (14)1995: 113-127.
12. Regan, D.H.: "Duties of Preservation", in Norton, B.G. (ed.): *The Preservation of Species*, Princeton, N.J.: Princeton University Press, 1986.

Sustainability and the liberal-communitarian debate[1]

Avner de-Shalit

The question I want to raise is this: which discourse is more promising in giving sustainability a chance: the individualistic-liberal, or the communitarian one? I want immediately to define the key concepts in this paper. First, the reason I use the term 'discourse' rather than 'theory' is that I want to examine the whole approach; the 'theory' is a concept limiting the relevance of the debate to the university or the academic world. By 'discourse' I mean the way ideas and theories percolate down from the ranks of academia and professional philosophy to society, and the way that society utilizes this information. I am aware that by so doing I run the risk of being less accurate, and hence less 'scientific', but I do think it's about time theorists theorize beyond theory. In other words, I want to look not only at ideas but also at their implications and applications.

Second, for many philosophers sustainability is a terrible concept. Thus it has been claimed there are so many different meanings for sustainable development, that it is a concept good for rhetoric only.[2] I am not sure I agree. It seems to me that although it may be extremely difficult to define sustainability, it is a concept upon whose use we all intuitively agree. For notice, that however sustainability is defined, in its common use by the general public it encompasses more than ecology and ethics. Rather, it is about society and the environment and the way social and political institutions treat the environment and society. In other words, when we talk about sustainability, we do not refer to ecology alone; in order for ecological systems to be sustainable, something should be done on the political level, and not merely with regard to the natural environment, but with regard to poverty, ignorance, and other matters which do not immediately seem to be related to the state of the environment. So we are looking at the way society uses its natural resources plus the ideology which accompanies this use. Sustainability is a matter of respectful and careful use. But if it is careful, for whom do we care? Or whom do we care about: only contemporary human beings? future generations as well? other species as well? And finally, is this use of natural resources detached from, or part of, a broad conception of social justice?

So now my opening question should read like this: what are the differences between individualistic liberalism and communitarianism in their approach to the environment and the relationships between the environment on the one hand and society and politics on the other? Do both recognize a link between the environment and other social goals? I shall start by looking on the positive side of liberalism and state that environmentalists owe a great deal to it. However, I shall then challenge individualistic liberalism as incompatible with sustaining sustainability in its political sense (which is the core of sustainability). I will go on to suggest that communitarianism is more in line with sustainability. Nevertheless, I shall argue that most interpretations of the relationship between the idea of 'community' and protecting the environment or any other environmental philosophy are misleading. Some of you might find this part of my talk controversial because I am critical of some of the ideas produced at this research institute. But at the end of the day, isn't that the reason why we have gathered here? Whatever the disputes are, it seems to me that we all share a deep belief in the crucial role of philosophy in protecting the environment, and that we all agree – as a great person once said – that the main point is not only to understand the world, but to change it as well.

Liberalism and the environment

Liberalism has been rejected by environmentalists for various reasons. Perhaps the most common position is that liberalism and the environment are incompatible because liberalism's most fundamental feature – the contract – excludes elements which cannot join it, such as trees, rocks, rivers, the Ozone layer, animals, and so on. John Rawls, for example, leaves this issue to metaphysics,[3] and although some have tried to modify his theory so that it generates an animal rights ethics or environmental policies[4], the critique holds true. The most contractarianism can do is generate obligations among humans *with regard* to the environment, and these can always be overridden by other obligations that humans have to each other. And yet, the fact that contract theories are difficult to apply to environmental matters should not imply that environmentalists should not have a strong respect for liberalism.

The first reason is that liberalism is at least partly responsible for the growth of environmental literacy. At this point I should distinguish between three concepts. The first is *environmental literacy*, which is a function of the information available to citizens about the environment. One is 'environmentally literate' when one knows about the Greenhouse effect, the Ozone layer, biodiversity, radioactive waste, and so on. However, one does not necessarily accept that this information has any particular moral implications. The second concept, *environmental awareness*, describes a state of mind in which one acknowledges that environmental matters affect our life to such a degree that they pose certain moral dilemmas, such as environmental justice and humans' responsibility to non-human animals. The third concept, *environmental consciousness* represents a deeper level of concern, wherein environmental matters are perceived as a political issue to be tackled on the political level rather than merely on the technological or even moral level.

Now liberalism, I have claimed, is partly responsible for the growth of environmental *literacy*. We all believe that informing citizens is highly important, but even if we assume that citizens are responsible and wish to know more, since there is so much information in the world today, people initially tend to be motivated to study something or get hold of a specific information only if they perceive its direct relevance to their lives. This is not a misanthropic observation, but rather a sympathetic and diagnostic comment. Among all the flashes of information one is exposed to, and between one "glide" in the network and another, wouldn't we all tend first to pick up those pieces of information which concern us? It is therefore crucial, from an environmentalist point of view, that people not only learn about biodiversity and air pollution, but that they realize that these issues affect them. In other words, there is a good chance for *environmental literacy* if it associates itself with *environmental awareness*. Indeed, environmental lawyers who are engaged in these matters see them as a threat to people's *rights* not to be made aware of unnecessary risk. Risk, the very opposite of sustainability, at least in its psychological dimensions, is a violation of one's rights. And it is in the context of such violations of rights that information about the environment is often conveyed, and not only in the Western world.[5]

In this sense liberalism has been a breeding ground for a flourishing environmental discourse, with particular reference to the tradition of defending the individual against the church, the state, and large-scale firms. Sustainability, then, is presented not only as defending the rights of other species, but also and first and foremost, as defending humans' rights. Liberalism's concern for human rights, then, helped to provide a *meta-language* in which information about the environment could be effectively conveyed.

However, liberalism is also responsible for the evolution of *environmental awareness*, due mainly to its strong attachment to anti-chauvinism. Our depletion of resources and the damage we cause to the environment can be described as *human* chauvinism towards non-human animals. (By chauvinism I mean failing without reason to respect others as equals, or the prejudiced support for one's narrow group). Indeed, just like Jeremy Bentham did years ago, many liberals have adopted the rationale that, in the absence of any clear and relevant difference between human and non-human animals, the latter should be equally respected. Perhaps the American legal theorist Christopher Stone paved the way this century for such an attitude, when he defended the trees' rights to stand in court.[6] So just as liberals in 19th century America moved from racism to moral universalism and liberals throughout the world in this century moved from male chauvinism in voting to equal enfranchisement for women, environmentalists today move from human chauvinism to a broader moral universalism, which extends Kantian or utilitarian ethics to include at least all sentient creatures.

Many people blame liberalism for being solely or principally responsible for an attitude to nature which exposes humans, not to mention non-human species, to unreasonable risks. How is it, they would ask, that you present liberalism as the source of the defence of the victims' rights? In reply I think we should distinguish between 19th and 20th century liberalism. In other words, I want to note a development of late 20th century liberalism: while liberals in the 19th century adopted the idea of progress wholeheartedly, and for the sake of progress were ready to turn a blind eye to its victims such as the working class, liberals in t he second half of our century have become much more egalitarian, to the extent that it is now claimed by Ronald Dworkin, a prominent liberal philosopher, that liberalism's most fundamental value is equality rather than liberty.[7] I think it can be argued that liberalism has become much more sensitive to progress's victims. Therefore, while maintaining its faith in progress, liberalism has developed a concern for the well-being of groups vulnerable to the side-effects of modernization, initially perceived only in terms of human beings (the working class, the elderly, etc.)[8] Contemporary liberals extend their recognition of vulnerability to children, ethnic, or gender groups. In view of the particular evils that modernity and progress have engendered, such as heavy pollution, toxic, nuclear and radioactive waste and misguided urban development, it is easy to see how the league of victims should come to include endangered species and other aspects of the non-human environment. I am not claiming that all liberals have adopted this attitude, but am making the more modest claim that it is feasible and reasonable to arrive at such a standpoint from contemporary liberalism.

Liberalism as a philosophy of government

Up to now I have explained how liberalism has contributed to the emergence of environmental literacy and awareness. But as I mentioned in my introduction, in order to fully grasp the meaning of sustainability we should attach priority to the political agenda. This, in turn, demands that people develop environmental consciousness, and it is at this point that liberalism has failed. Why so?

The first point that comes to mind is the role of market forces. Individuals' economic behaviour, which is motivated by profit-seeking is held to be both inefficient and inequitable in coping with ecology, and therefore in supplying the rationale for sustainable ecological systems. Moreover, as Finn Arler so eloquently explains, the market is based on what he calls *"utility friendship"*, in which the main criterion for allocation of goods is that of entitlement. So those who have the means also have the right to buy, and justice *"resides alone in fair bargaining"*. No obligation to help the needy, whether those are defined in economic or environmental terms, is included. So this utility friendship cannot form the basis of international and intergenerational justice, or in other words, sustainability.[9]

I wish, however, to tackle the issue of liberalism by going beyond the technicalities and the principles of the market. I want to address the moral foundations of liberal politics, leading, in turn, to the issue of neutral governing vs. the politics of the good. In other words, I want to address sustainability in its full sense.

The problem starts when we move from discussing sustainability morally to a politics of sustainability, and it becomes necessary to convince those who oppose it, e.g. insensitive developers or vivisectionists, of the need for environmental policies, legislation and institutions. The individualistic-liberal conception of politics is based on values of neutrality and minimal state intervention, and a view of politics as an aggregation of autonomous decisions made by individuals. Will this be sufficient, considering that the two sides – e.g. developers and preservationists – have contradictory agendas?

Liberals often claim that neutral politics, the policy of following the 'invisible hand', or simply the aggregation of private, individual wills, yields the "best" results. This claim has two faces; first, it is argued that such policies increase utilities and are rational. But second, and no less important, it is claimed that such an attitude to politics guarantees maximum liberty by by-passing questions of what constitutes the good life. But sustainability is precisely such an idea of the good life. It has implications for the public and, although some would doubt this, private aspects of our life. So how can we attain sustainability if we are to ignore ideas of the good such as sustainability?

At this point I should substantiate my claim that sustainability is an idea of the good. I'll do this by referring to an argument I have already made elsewhere, as a rejection of Dworkinian theory of obligations to future generations. Suppose, then, that a liberal neutralist politician thinks it is necessary to preserve a certain piece of land in order to attain sustainability. This politician now wants to present a neutral argument for preservation. In the absence of a clear majority supporting this policy, the only argument our politician can put forward is that we contemporaries do not want to deny future generations the enjoyment of this good, not because this good is part of our idea of the good, but rather because if we destroy it they will no longer be in a position to make a neutral choice among several alternatives, now available.[10]

However, this way of reasoning is misleading. If you and I are choosing between P objects and I assert that we should also consider object $P+1$, you will probably ask me why. I cannot reply 'because it is there', because then you may wonder why I did not consider object $(P+2)$. I therefore have no other choice than to explain why $(P+1)$ is especially desirable, significant, important, and so forth. In other words, either we leave future generations *everything*, including for example nuclear bombs (because we feel obliged to make every effort to enable future generations to practice choice), or we face the fact that preservation is founded upon notions of the good. We should preserve trees rather than nuclear weapons; the former are good, the latter are bad. Discussing preservation (as part of sustainability), then, is done from the perspective of one's conception of the good.

At this point some liberals could try and defend their position by claiming that even within liberalism there is a school of thought that attaches importance to participatory rights, social rights, and so on. Although this is true, it cannot challenge my critique because both traditions within liberalism end up supporting neutrality. There are liberals who start from neutrality (because of minimalism, that is the idea that since there is a variety of beliefs, values, and ideas of the good, the state will never be able to take side without oppressive measures, and therefore politics should stay out of the moral debate) and construct an egalitarian theory on this basis; and there are liberals who, admittedly, start from a broad egalitarian principle (e.g. the equal inherent worth of human beings), which may lead them to support social rights, but they too end up backing a neutral position, arguing that state neutrality reflects the idea of the equal worth of human beings. They will therefore be reluctant to discuss environmental policies in terms of the good.

To rephrase my argument; liberalism has contributed to the emergence of environmental awareness, according to which non-human entities may have rights, interests, or their own, intrinsic good. In order to defend such a position, an environmental argument will tend to rest on a theory of value. Liberalism as a philosophy of governing, however, is about choice and autonomy. Those who are autonomous to make choices are human beings. So the only thing that legitimately limits one's choice is harming other *human beings'* rights. Other entities, by contrast, do not count. This, as I say, contradicts environmental awareness (and the understanding that other species count as well and that they are part of the good in this world) and any such theory of value (i.e., a theory that relates to the idea of the good). The liberal theory of politics, therefore, is inconsistent with the liberal theory of knowledge (environmental literacy) and the liberal theory of morality (environmental awareness). There is, then, a need for a different theory of politics.

Community and the environment

Now let me depart for a moment from this dry analytical style. Let me ask you to imagine several models of community. The first is a multi-species community. It is probably an "alternative" commune somewhere in India or Brazil, mostly inhabited by European "freaks". They believe that humans and nature form one large community, based on metaphysical assumptions. Nature, they claim, is part of humans and being human is being part of a larger community consisting of many species. Only by realizing this can one fully comprehend the meaning of being human. From time to time these people refer to conservation practices of indigenous peoples in the Third World, but to a large extent, their approach derives from Aldo Leopold's famous book, which every member of this commune has (but does not *own*) by his or her bed, namely *A Sand County Almanac*.[11] Land is a community, argued Leopold, and morality is nothing but sustaining balance within this community, by changing *Homo Sapiens* from conqueror, or at least owner of this community, into a member of it. This notion of a single community, in which a single code of ethics governs human behaviour towards both human and non-human entities is also reflected in these people's quest for a new ethics, i.e., monolithic, 'environmental' ethics. This is to say that as far as they are concerned, environmental ethics is not about humans' relationships to the 'environment'; rather it is ethics itself. In other words, environmental ethics substitutes ethics.

The second community is another "alternative" village, somewhere in West Europe. It consists of animal welfarists who believe that since humans affect and are equally affected by nature, they are no more than single-species components of large ecosystems, just like any other species. Thus, these villagers claim, there are no moral grounds for differential, privileged treatment or respect for humans. Accordingly, justice should include animals, and reasoning about justice should include animals' perspective. These people often refer to books such as Tim Hayward's recent, and may I add very interesting book, *Ecological Thought*,[12] which calls for humans to get rid of their superstitious beliefs about nature and thereby become more able to understand nature. These animal welfarists distinguish themselves from the former, community of preservationists, in that they care for the individual animal rather than for balancing ecosystems. They often invite anti-vivisectionists such as Peter Singer and people from the Council for Animal Welfare or the International Society for Animal Rights to talk to them. They argue in favour of minimizing pain and misery, often on the basis of sentimental approaches which assign a key role to empathy. It is clear to them, however, that empathy, which is the power to enter into the feelings of others, assumes a certain affinity between the feelings of humans and animals; in other words it assumes a strong sense of community. Thus this community is constructed on the basis of an egalitarian interpretation of biology.

On the face of it, the third model of community is nothing to write home about: not only is it not related to any exotic location, but it is not related to any specific location at all; it consists of a much larger number of people, and of people only. Its members – despite being eager to help non-human animals – do not see a way of using the term community to describe relationships with animals. And yet, it is a very strong community with specific goals and intense political and moral deliberation. People often participate and are encouraged by their fellow members to do so through a variety of means of activity or communication. This community has established and built institutions which enable people to express their thoughts and ideas regarding their public life. As for non-human animals, however, they usually enjoy a lot of sympathy, but not necessarily empathy.

I would now like to analyze these models of community and argue that the first two are inconsistent notions of community and are not reasonable. However, I shall defend the latter as a very promising way of dealing with sustainability and with the environment, although at the moment it is far from being a satisfactory mechanism for environment-friendly and sustainable policies.

The first claim I would like to examine is that human beings are not different from any other species. This is a metaphysical claim, and as many other such claims, it is a matter of belief. Thus I have very little to say about it. However, some of the implications drawn from such a standpoint do not make sense. The fact that humans are natural cannot imply that the organization of human politics should be more 'natural', whatever this means. It has, for example, been argued, that the principal features of the natural world and the political conclusions that can be drawn from them are: diversity – toleration, stability and democracy; interdependence – equality; longevity – tradition.[13] Now, it is not quite clear from this whether we should start by looking at nature and realizing that it combines diversity with sustainability, hence conclude that we should follow this successful line, or whether we should start by thinking that we should be tolerant and live in pluralist communities, only then to find "proof" for this standpoint in nature. Whichever of the two is the attitude, it is wrong. There is no reason why we should derive harmony rather than some other characteristic from nature. It seems arbitrary, or even dogmatic, to perceive harmony rather than, for example, conflict or competition in nature. Similarly, there is no reason why one should find balance and stability in nature rather than cruelty, constant change, and misery. Nature encompasses so many phenomena that it is possible to "find" almost anything in it. Moreover, what we find in nature may be nothing but a reflection, or even projection of human values (rather than authentic, 'natural' ones) because nature itself does not hold, nor is it governed by, values.

Which leads me to the thorny idea of deriving rules of conduct for one arena (relationships between humans, i.e. politics) from another (relationships between species, i.e. nature). Although we often have an intuition that things are "just the same" we should be cautious about importing paradigms from one sphere to another. In this case we are asked not only to transpose nature onto human beings, but also onto politics, and herein lies the problem. Nature is natural, whereas politics is artificial and cultural. Show me one other species that votes, demonstrates and organizes itself into parties and I will withdraw.[14]

Still, an anarchist could claim that this only proves her point: if there is no politics in nature, then such should be the case with us humans. Let's organize, then, in natural communities. Let's return to nature. However, there is a further complication here; some elements of and processes in nature are good for nature, but bad in the context of society. In other words, if there is a community in nature, such processes as these sustain it, but would damage or be considered immoral in a human community. Think, for example, about a parasite, or about killing, which is part of everyday life in nature, but is either immoral (murder) or irregular (war) in social life. Or think of early death which is sometimes required by nature in order to renew life and sustain the ecosystem, but in social life is a tragedy and might destroy systems such as the family.

So, I think, the multi-species community is not a coherent concept of community. Moreover, it contradicts a basic intuition that many of us share. Consider, for example, a case in which I witness a dog chasing another one. I should stop it from harming the other dog, and if it does, I should take the harmed dog to the vet or to its owner. However, if I walk in the mountains and see a bird attacked by a bird of prey I should not intervene because this is "nature's way": perhaps the death of this bird helps maintain equilibrium in the ecosystem. So if indeed my obligations to domestic animals are stronger than to wild animals, it seems reasonable to claim that there are different levels of human obligations to animals and therefore there are different communities at stake. If so, then, why is it unreasonable to claim that a superordinate level of human obligations exists, namely to other humans?

Is it wrong, then, to talk of 'community' in the context of nature? It may not be, if we accept that unlike the community of human beings (to which I shall soon turn), 'community' in nature is a metaphor, and should remain as such. That the two – community in nature and the real, human community – are distinct from each other is revealed in the connotations attached to these terms. Human communities remind us of *mutual responsibility* whereas the natural 'community' reminds us of *interdependence*. Human communities implies being *tolerant to the other*, whereas community in nature is not only the definition of *borders and boundaries*, but also their defence. This is, I admit, only a matter of rhetoric, images, and so on, but it does point to a fundamental difference between nature and human politics.

However, if we return to the various models of community, we are still left with the one that on the face of it looked rather pale and unattractive. I have in mind the larger community which consists of humans alone. But why is regarding social and political life as being exercised within human *communities* likely to sustain sustainability? I shall not go into too much detail about this model of community because several answers can be found in the literature about environmentalism and political theory.[15] However, I shall say a few words beyond the argument there. Sustainability, I claimed earlier, is an idea of the good. Environmentalism is the manifestation of the good life, according to environmentalists. So when environmentalists become engaged in the decision-making process upon environmental policies, they want these policies to be moral, that is to grasp their moral content. Now, these policies will be moral not because they reflect a consensus or because they allow us to coexist; rather it is the case that after long and profound deliberation, the community has realized that *this* act is the just one with regard to the environment because it ties in with the relevant principles of morality. Admittedly, the 'relevant principles of morality' is an expression pointing to the existence of a moral community. But this does not have to be a multi-species community. Asserting that a certain community should care for X does not imply that X is part of this community; it may simply imply that X is part of this community's conception of the good.

Now, in order for us to reach the stage of 'relevant principles of morality' we must first have communitarian politics; communitarian in the sense that one discusses and questions one's beliefs and principles through a sympathetic consideration of others' ideas, and society decides upon the right policy according to the policy's moral status rather than according to its procedural value, i.e., its ability to allow us to live together. If environmentalism is to win and sustainability to gain respect, it is only through profound deliberation of the morality of these standpoints. No other way is possible: enforcing people to accept sustainable policies is stupid because it is impossible in the contemporary political atmosphere which highly respects – and rightly so – individual liberties, and because it will be counter-productive. Going for the policy which allows us to live together is also fruitless, not only because instead of discussing our obligations to the environment we shall be dealing with obligations to each other, and not only because there are 'bad guys', i.e. those who through private profit motives oppose environmental policies, but also because the 'good guys' are likely to differ in their view as to what constitutes better understanding of our duties to the environment. Imagine, for example, a case where there used to be an ecosystem E1 several years ago, but due to human activities there is now ecosystem E2, which includes a variety of species. Should we preserve and protect E2 or should we intervene and renew the original E1 to compensate for human intervention in the past? To make things even more difficult, imagine that several species may become endangered if they are not brought back to this area and protected as part of E1, whereas other species, currently living in E2 may become endangered if the former species are brought back.[16]

I personally don't know whether I have an answer to this question. But this is beyond the scope of this paper. What I want to emphasize is that the only way to come up with an answer to such questions which is not a technical compromise leading to the neglect of our duties to the environment, is through debate, by trying to convince and being open to the arguments of others. In short, it is communitarian politics.

Concluding remarks

At this point an argument can be raised, according to which this model of human community is not satisfactory. "Look around you", the claim would go, "Do you see sustainable policies? Of course you do not. Well, then, what we need is a complete change of mind; a new paradigm of thinking."

My reply is twofold. First, when I look around me all I see is the market, market forces, and Capitalism. The reason we do not have sustainable policies is not too much of the communitarian model but too little of it. There is too much of individualistic materialism, while the state is hiding behind the idea of the 'neutral state'. A question may be raised here, namely what if the community decides for Capitalism. But I think that this is not a real option, if we have in mind a genuine community. Capitalism and material individualism undermine community, because they regard competitive individualism and economic efficiency as very important values in their system. Thus the community is pushed aside in order to allow the market and the individual flourish. Capitalism must, again, rely on neutrality and minimal state intervention, which are, I believe, the opposite of community. Thus a genuine community will not decide in favour of Capitalism.[17]

Second, I wish to point to what I call the "tragedy of environmentalism". In order to convince through communitarian politics, people must be engaged in the debate and the right conceptions of the good put forward. However, environmental theorists, philosophers and activists have alienated themselves from this community. They have developed their own language – biocentrism – which may be perfectly all right (again, if it does not rest on the multi-species community), except for the problem that the majority of the population does not speak and does not understand this language. By being obsessed with questions of meta-ethics, intrinsic value, and proving that there is a multi-species community, environmental theorists have neglected the main-stream political and moral discourse. Thus the latter has been abandoned, only to be occupied by developers, Capitalists, and their ideology. The model of human community is fruitful, but it all depends on the idea of sustainability being supported – within the communitarian discourse – by a larger part of those who want to defend it.

To conclude, I have claimed that liberalism has played a key role in the emergence of environmental awareness and in the diffusion of environmental literacy. Therefore the criticism – in fact, the total rejection – of liberalism by many environmentalists, is far-fetched and unwarranted, not to say unwise. However, this is true for liberalism as a set of values; the liberal values. But liberalism is also a political system. It is at this point that liberalism is not sufficient to ensure the emergence of an environmental consciousness, and hence the need for a shift to the politics of community and of the good. To date, the way most environmentalists have discussed community is too simplistic and conceptually faulty. This, however, can be rectified if nature is left to be natural, and community returns to be the political and human framework within which environmental policies are discussed. Such a discourse is more promising in giving sustainability a chance.

Notes

1. In writing several drafts of this paper I benefited from many comments. In particular I would like to thank Finn Arler, Maurie Cohen, Lord Dahrendorf, Stephen Holmes, David Miller, Bryan Norton, John O'Neill, Darrell Posey, and Gayil Talshir.
2. See Andrew Dobson's important article: "Environmental Sustainabilities: An Analysis and a Typology", *Environmental Politics*, Vol. 5 (3), 1996, pp. 401-429.
3. Rawls, John: *A Theory of Justice*, Oxford: Oxford University Press, 1973, p. 512.
4. Singer, B.: "An extension of Rawls' theory of justice to environmental ethics", *Environmental Ethics*, Vol. 10, 1988, pp. 217-32; Taylor, R.: "The environmental implications of liberalism", *Critical Review*, Vol. 6, 1992, pp. 265-82.
5. In fact, lawyers have been arguing two distinguished claims. First, that risk itself is a violation of one's rights; this, however, may be controversial: do we have a right not to be under any risk? But the second claim is more intuitive: not informing us or disinforming us about a certain risk is a violation of our rights.
6. Stone, C.: "Should trees have standing?", *Southern California Law Review*, 45, 1972.
7. See Dworkin, R.: "Liberalism" in his *A Matter of Principle*, Oxford: Clarendon Press, 1986.
8. See Goodin, R.: *Protecting the Vulnerable*, Chicago: Chicago University Press, 1985.
9. Arler, F.: "Justice in the air: Energy policy, greenhouse effect, and the question of global justice", *Human Ecology Review*, Vol. 2, 1995, p. 49.
10. Dworkin, R.: *A Matter of Principle*, Oxford: Clarendon Press, 1986, p. 202.
11. Leopold, A.: *A Sand County Almanac*, Oxford: Oxford University Press, 1987 [1949].
12. Hayward, T.: *Ecological Thought*, Polity Press, 1994. I should add that there are several aspects in Hayward's theory which these people reject.
13. See Dobson, A.: *Green Political Thought*, London: Harper and Collins Academics, 1990, p. 24.
14. I am aware of the post-modern claim that nature is also 'cultural'. But I see no reason to refute this here, since the whole project of environmental philosophy is to protect nature as natural, and my point is that *this* cannot live in harmony with the idea of deriving politics from nature.
15. See O'Neill, J.: *Ecology, Policy and Politics: Human Well-Being and the Natural World*, London: Routledge 1993; and de-Shalit, A.: *Why Posterity Matters*, London: Routledge, 1995.
16. Notice that Ingeborg Svennevig is discussing a parallel dilemma with regards to the local farmers' interests, in her 'The social significance of nature protection in the Wadden Sea area', paper presented at the international conference 'Cross-cultural protection of nature and the environment', Odense, May 1997.
17. This is not to say that the community will not decide in favour of a welfare state which includes modified market economics. I elaborate on this in the final chapter of my book, *The Environment Between Theory and Practice* (forthcoming).

The Good Life below the snow-line: pluralism, community and narrative

John O'Neill

The topic and title of this paper were prompted by reading a few stanzas from a poem of W.H. Auden – "Letter to Lord Byron". Much in that poem represents a criticism of what Auden calls the "*excessive love for the non-human faces*" and as such it anticipates some recent criticisms of the anti-humanism of deep greens. However, this theme is not my central concern me here. The stanzas that prompted the thoughts that follow were these:

> The mountain-snob is a Wordsworthian fruit;
> He tears his clothes and doesn't shave his chin,
> He wears a very pretty little boot,
> He chooses the least comfortable inn;
> A mountain railway is a deadly sin;
> His strength, of course, is as the strength of ten men
> He calls all those who live in cities wen-men.
>
> I'm not a spoil-sport, I would never wish
> To interfere with anybody's pleasures;
> By all means climb, or hunt, or even fish,
> All human hearts have ugly little treasures;
> But think it time to take repressive measures
> When someone says, adopting the 'I know' line
> The Good Life is confined above the snow-line.
>
> Besides, I'm very fond of mountains, too;
> I like to travel through them in a car;
> I like a house that's got a sweeping view;
> I like to walk, but not to walk too far.

> I also like green plains where cattle are,
> And trees and rivers, and shall always quarrel
> With those who think that rivers are immoral.[1]

Now I should confess at the outset that I am one of Auden's mountain snobs, although I question the degree to which I'm the fruit of Wordsworth – poets, like the rest of us, are apt to exaggerate their own influence. But in most other respects, strength aside, I fit the bill. I don't shave; I enjoy climbing; I take perverse pleasure in discomfort; I find great happiness above the snow-line. On the other hand, car rides in mountains induce little but travel sickness. My tastes about mountains differ from those of Auden.

However, there are two themes in Auden's poem which deserve serious consideration and which have been developed in more prosaic terms in more recent criticism of environmentalism. The first is the extent to which much in modern environmentalism might be taken to be a local vision. The second is the potentially authoritarian way this local vision is imposed on others. A particular conception of the good life, according to which the good life has to include a life above the snow-line, in a wilderness devoid of human faces, is presented as a universal vision of the best life for humans. The worry is that much in modern environmentalism is simply the attempt to impose that local conception of the good life on others. Correspondingly there is sometimes a failure in modern environmentalism to appreciate the variety of ways in which different human relationships to the natural environment can be lived in the plains. That worry is of particular significance for Aristotelian positions such as that I have defended elsewhere[2] according to which a proper relationship to the natural world is taken to be a constitutive component of the good life. The Aristotelian and communitarian view that public institutions should aim at the good life has long been criticized on the grounds that it makes for an authoritarian politics that is incompatible with the pluralistic nature of modern society.[3]

Authoritarian environmentalism?

To capture the very real potential authoritarian consequences of environmentalism I turn to a second critical commentary on the Wordsworthian vision of nature, a 1929 essay of Aldous Huxley, 'Wordsworth in the Tropics'.[4] Huxley opens with remarks on both the localism and influence of the Wordsworthian vision: "*In the neighbourhood of latitude fifty north, and for the last hundred years or thereabouts, it has been an axiom that Nature is divine and morally uplifting. For good Wordsworthians – and most serious-minded people are now Wordswor-*

thians either by direct inspiration or at second hand – a walk in the country is the equivalent of going to church, a tour of Westmorland is as good as a pilgrimage to Jerusalem."[5]

Now like Auden, Huxley here exaggerates I think the influence of poets – but this I leave aside and for the moment use the term 'Wordsworthian'. The central claim of Huxley's essay is that the Wordsworthian vision would not travel well. It is possible only in the domesticated environments of Europe, not in the non-temperate zones. *"The Wordsworthian who exports this pantheistic worship of nature to the tropics is liable to have his religious conviction somewhat rudely disturbed. Nature under a vertical sun, and nourished by the equatorial rains, is not at all like that chaste, mild deity who presides over…the prettiness, the cosy sublimities of the Lake District… There is something in what, for lack of a better word, we must call the character of great forests…which is foreign, appalling, fundamentally and utterly inimical to intruding man."*[6]

For all Huxley's perceptive remarks in his essay on the conditions for certain kinds of appreciation of nature, his claim here is thoroughly mistaken. In the first place, subsequent developments have shown the Wordsworthian vision to be capable of travelling much better than Huxley assumes. It has translated itself well into the tropics. Modern nature lovers no longer confine themselves to the Lake District or the Alps. The trip to tropics has become part of the Wordsworthian pilgrimage and a whole tourist industry exists dedicated to the transport believers to non-temperate zones. Nor is this export of Wordsworthian vision a recent development. Indeed, Huxley's own account of alien nature of the tropics is part of the export industry. His own view is much closer to the Wordsworthian vision than he assumes. And this takes me to a second point about Huxley's characterization of the tropics as places of wilderness which are 'fundamentally and utterly inimical to intruding man' and which are in stark contrast with the domesticated environments of Europe – that characterization is drawn in European colours. The account renders invisible all those indigenous persons, women and men, for whom the tropics were not and are not an alien and impenetrable wilderness, but their home.

And here I turn to the potentially intolerant and authoritarian dimensions of the environmentalist vision. Huxley's Wordsworthian is not only doing well in the tropics, but the nature itself is being remoulded to confirm with his image of it as a place in which humans have no place. Nature parks are created and legitimized through the wilderness model of nature conservation which puts considerable emphasis on the values of wilderness, understood as nature untouched by humans, and of 'ecological integrity', understood as the integrity of ecological systems from human interference. This wilderness model developed historically through that image of tropical nature invoked by Huxley as an unspoilt wilderness that contrasts with the domesticated environments of Europe.[7] This shaping of non-temperate nature in conformity with the European image of a wilderness is achieved through the coercive exclusion of those whose home the wilderness is. Thus the development of conservation parks in the third world through the eviction of the indigenous populations that had previously lived there. Consider the fate of some of the Masai who have been excluded from national parks across Kenya and Tanzania.[8] Attempts to evict indigenous populations from the Kalahari reveal the influence of the same wilderness model: "*Under Botswana land use plans, all national parks have to be free of human and domestic animals.*"[9] Nor is the policy of enforced eviction confined to Africa. Similar stories are to be found in Asia where the same alliance of local elites and international conservation bodies has lead to similar pressures to evict indigenous populations from their traditional lands. In India, the development of wildlife parks has lead to a series of conflicts with indigenous populations. Thus, in the Nagarhole National Park, there are moves from the Karnataka Forest Department to remove 6000 tribal people from their forests on the grounds that they compete with tigers for game. The move is supported by international conservation bodies. Hence the remark of one of experts for the Wildlife Conservation Society – "*relocating tribal or traditional people who live in these protected area is the single most important step towards conservation.*"[10] The wilderness model fails to acknowledge the ways parks are not wilderness but a home for its native inhabitants, the degree to which the landscapes and ecology of the 'wilderness' were themselves the result of human pastoral and agricultural activity, and the cultural significance of particular landscapes, flora and fauna to the local populations.

The conflicts between the Wordsworthian and indigenous populations are not confined to the tropics. Similar conflicts are played out in a less dramatic way back at home in the more traditional places of pilgrimage. These too have been shaped to produce a landscape that conforms to a particular Romantic aesthetic. Consider for example conflicts in the Dales National Park in Yorkshire.[11] Farmers on the one hand and landscape planners and conservationists on the other have different perceptions of what constitutes a good environment of which some are self conscious. Farmers' perceptions are often husbandry based. Hence the comment of a farmer in the Yorkshire Dales: "*A farmer will look at someone else's farm and could tell whether it was well farmed or not. They wouldn't look at the view and think 'What a good view!'*"[12] Given that perspective the wildness loved of the Wordsworthian can be seen as a defect. "*If a piece of land's conserved, it tends to get overgrown, it gets brown. I suppose people from off will tend to look at that and admire its tones, in autumn sort of thing. Or golden spring, or whatever. But a farmer will look at it and think – it's overgrown.*"[13] The farmer will sometimes look upon the land in a way that is different from both the visitor admiring the landscape – although farmers are aware of the landscape – and the nature conservationist. Attempts to fence off nature and allow it to grow wild is met with disapproval: it represents a 'mess'.[14] Hence, the resistance felt by some farmers to the way that the authorities who represent conservationists and landscapers – outsiders who aim to mould the environment in ways that are alien to their own husbandry based conceptions. Correspondingly there is the articulated threat to a community founded upon farming being transformed into a museum exhibit to conform to some idealized Romantic image of how the countryside should look: *"National Parks, English Nature, they'll finish up with all the farmers running about in smocks, like museum curators. That's not a community. We have a community which is a working community..."*[15] The worry here is that a particular conception of the way nature and landscape ought to be is being imposed from the outside on those who live and work in an environment, for whom nature is not primarily an object of scientific interest or aesthetic contemplation, but something with which one has a working relationship.

The good life, pluralism and relativism

Thus goes the case for the prosecution of environmentalism: environmentalism exhibits a worrying authoritarianism and intolerance which is blind to the ways a good life can be lived below the snow-line and which imposes its vision on populations who have a different relationship to nature – notably those for whom the natural world is the scene of their agricultural and pastoral lives. I have tried to state the case as robustly as possible because it appears to have particular power against one particular philosophical position about environmental value that I have defended myself.[16]

The position which it threatens is that which defends environmental concern on broadly Aristotelian grounds that a proper relation with the natural world is a constitutive component of the good life.[17] To take another statement of the position, consider Nussbaum's claim that *"a creature who did not care in any way for the wonder and beauty of the natural world"*[18] would lack an element that is constitutive of what it is to be human. The case for the objector would run that the authoritarianism of the environmentalist merely exhibits just what is wrong with Aristotelian account of the public life. The classical view of politics according to which *"the end and purpose of the polis is the good life"*[19] is incompatible with the pluralism of modern life. Specifically an environmentalism which is founded upon the premise that some particular form of relationship to nature is a component of the good life is founded upon a failure to recognize the variety of ways a good life can be lead without any the particular relationship to nature proffered as best. There are a plurality of different ways a good life can be lead below the snow-line involving a relation to nature which is very different from that advocated by many of the defenders of environmentalism who call upon some universal shared appeal to the 'wonder and beauty' of the natural world. There are individuals who live perfectly good flourishing lives below the snow line, for whom nature is a place for their working life, or who, like Auden, love the life of the city and are content with the nature of the garden or park or the view from the car. The objection runs that, in so far as the modern green movement does involve a kind of Aristotelian perfectionism, it issues in just the kind of authoritarianism outlined in the last section. It leads inevitably to the coercive imposition of a particular vision of the good life on others who do not share it, whether this be within the confines of a particular state or through the missionary activities of its proponents across states. Hence, the common rejection of the very idea of appeal to a shared and universal conception of the good life by both liberals and postmoderns.

Is there a response that can be made on behalf of the Aristotelian position? I think there is and the rest of the paper will be concerned to develop such a response. I start however by indicating why I think the Aristotelian position is worth defending. Very briefly it is because I believe that neither a thin theory of the good of the kind that a Rawlsian liberalism offers nor the forms of relativism about the good that are popular amongst postmoderns can provide an adequate basis for criticism of the ways in which life in modern capitalist societies disrupts a proper relation to the natural world, and that a broadly Aristotelian position can. And here I take issue with the claim that relation to nature that Auden and Huxley characterize as Wordsworthian is just the product of an English Lakeland poet. It was a response also to the quality of life experienced in the new industrial cities. That response animated, for example, the mass rambling movements of the working class in Northern England who struggled for access to the countryside. At same time as Guha and Martinez-Alier note there are a variety of third world environmentalisms of the poor, which have their foundations in the degradation of the quality of their environment, not just in terms of basic life support functions – life in the sense of mere survival can be led amidst environmental squalor – but also the quality of life experiences and sociality it makes possible.[20] There is a need to hold onto a conception of human flourishing to found in these criticism – not least because the worst of conditions is that in which such loss is no longer experienced as such. I cannot see how a sound form of criticism can be launched without appeal to a thick theory of the good with some universalism about the conditions of human flourishing and the requirement of an adequate environment as part of those conditions.²¹

What response can be made then on behalf the Aristotelian position? Now in one sense a response to the specific cases discussed in the first section of this paper is not difficult: one does not need a highly complex theory of well-being to state what is wrong with using coercion to force large groups of people out of their homes and onto scrubland and leaving there to suffer malnutrition and disease. Moreover there are clear problems with the wilderness vision of nature which I have already touched upon. However, such specific responses do not touch the more general worry about the compatibility of the Aristotelian perspective with the pluralism of modern societies. In the rest of this section I give some general considerations from within an Aristotelian perspective for both defending a relationship with nature as a constitutive component of the good life for humans while allowing a plurality in the ways this can be lived. I aim to defend a pluralism without relativism. The argument will be in three stages: vagueness; plurality and community; and narrative.

Plurality and Vagueness

The existence of a plurality of different ways in which human life is lead and the problem this poses for claims to universality are not new. So just as Aristotle appeals to the kinship of humans experienced in their travels,²² so other classical Greek texts had pointed to the strangeness experienced. Thus take the following illustration recounted by the greater teller of strange tales for distant lands, Herodotus: *"When Darius was king of the Persian empire, he summoned the Greeks who were at his court and asked them how much money it would take for them to eat the corpses of their fathers. They responded they would not do it for any price. Afterwards, Darius summoned some Indians called Kallatiai who do eat their parents and asked in the presence of the Greeks...for what price they would agree to cremate their dead fathers. They cried out loudly and told him to keep still. That is what customs are, and I think Pindar was right when he wrote that custom (nomos) is king of all."*²³ Herodotus is often taken here to be illustrating the variability of standards of conduct according to the customs of a culture against those who held that there existed universal standard grounded in nature (physis). What is horrific for the Greek is the norm for the Kallatiai and vice versa. Custom is king of all.

This appeal to variability in responses to found a thorough going relativism is however less convincing than first looks suggest. The differences in question, like any other pair of significant differences, depend on the existence of shared understandings. Darius's elicitation of protests relies on understandings that are shared by the two groups, the Greeks and the Kallatiai, concerning the existence of special ties to particular others, more specifically to ones dead parents. The two groups express those relationships in different ways, one by cremation, the other by eating, but that the differences make sense against the background of a notion of the respectful treatment of the dead. One part of cross-cultural interchange, where this can issue in understanding, lies precisely in the interpretation of the actions of others that makes them not only understandable but expressive of relations and attitudes that we share at a deeper level. What in our culture would be shocking, eating your mum and dad, in another turns out to be expressive of a relation of respect to them. The story of Darius would be very different if, as we tried to unpack the act of eating one's parents, it turned out to express an attitude of deep hatred between generations.

Similar points apply to many apparent differences in cross cultural relationships with nature. Many differences are premised on shared understandings: one can be both critical of hunting in one's own culture or in the Roman colosseum, where it expresses an attitude of gratuitous enjoyment in cruelty, while accepting that hunting among the Eskimo might involve the expression an attitude of respect for fellow creatures in a context of common neediness. The same activity can have different meanings, and it is only when we understand it that a judgement can be passed. Correspondingly one failure of respect in an intercultural context lies in a blindness to the specific meaning relationships, objects and places have to others. What is absent in the case of the wilderness model is the acknowledgement that these are the homes of others, not a wilderness. However strange and alien the non-temperate regions might be to a European like Huxley that is not what it is like for the indigenous.

Nussbaum makes a similar point in noting that any defensible theory of the human good should be thick but 'vague in a good sense': it will allow that are a variety of different specifications of ways in which humans can lead a flourishing life, whilst offering an account of the powers, capacities and needs that make us the kind of being we are which does delimit the target of what a flourishing life can be.[24] The difficulty that a universalist conception of human flourishing has lies in steering a course between vacuity on the one hand and an implausibly narrow specification of flourishing on the other. However, that is a difficulty not an impossibility: a course can be steered and as such it still has real work to do. For example, to take a standard and I think true claim that is found in naturalistic theories of the human good that go back to Aristotle, as humans we are beings that need intimate relations to particular others. That claim does not determine some particular form such relations have to take – it allows of variability. The relations can take a variety of specific forms in different social and cultural settings. The Darius story illustrates just how dramatic that variability can be: however, it leaves room for criticism of a society like our own, in which those who are old and without wealth are excluded from ties of affiliation with others. Similar point apply to nature. One can as Soper puts it in summarizing Raymond Williams *"recognize both the permanence of the need for the countryside and the cultural relativity of its expression."*[25] In recognizing the former gives one critical ammunition against social forms that deny the satisfaction of the need for access to green spaces; in recognizing the latter one can accept that different kinds of landscape can satisfy that need.

Plural goods and social union

The appeal to vagueness does not however adequately respond to Auden's objection to the claim that a proper relation with the natural world is a constitutive component of the good life. Surely it might be argued there are those, the wen-men, who live flourishing lives with little or no contact with the nature – for whom the need for the countryside is not felt. Within our own culture, there are individuals who live perfectly good flourishing lives not just below the snow line, but with little or no contact with nature, who love the life of the city, dusty library rooms, art galleries, conversations in restaurants and only with some reluctance can be goaded into the occasional excursion into the garden or park, and does so merely to humour their outdoor companions. It is a form of life I might not share but one I have met and can appreciate. There are different forms that a flourishing life can be lead which realize quite different kinds of good.

This observation is compatible with the inclusivist version of the Aristotelian position, which accepts that the goods that go up to make a flourishing life are irreducibly plural. A flourishing human life contains a variety of intrinsic goods which cannot be reduced one to another.[26] Given that plurality and the fact that we are not angels but beings that work within limits of capacity, time and resources, we face practical choices in their pursuit of such goods. Some groups of activity will make up a form of life that is incompatible with others: one may not be able to lead a life of contemplation and a life of action. The pursuit of such activities may conflict with the calls of other relationships. Such conflicts give rise to the practical dilemmas of individual lives. There is no algorithmic procedure in such cases – rational choice is made on the basis of practical judgement. Choices may be more or less difficult. However, the limitations of individual lives force such choices upon us. And we can expect different individuals to realize different goods in their lives. The wen-men may lead culturally fulfilling lives in one way, others lead it in another.

What is true, however, a good society will be one in which there is expression of care for the wonder and beauty of nature in the lives of some of its members, along side those who pursue other goods. The recognition of this point provides the foundation however of Aristotle's defence of plurality within the polis against Plato. The goods realized by the polis, while bounded, are wider than those any individual member can realize. I cannot realize excellence in several fields – *we* can. This greater range of the goods realizable within the polis enhances the lives of its members. Through my relations with others, I can have a vicarious interest in these goods. Consider the case of friendship: To have friends with a diversity of interests and pursuits extends me. In caring for the good of my friends, I care for the success of the projects in which they are involved, for their realization of excellence in the activities they pursue. A friend *"shares his friend's distress and enjoyment."*[27] Thus in friendship the ends of another become one's own. Hence, while I may not be involved in such activities, I have a vicarious interest in the achievement of goods within them. My concerns are extended by those around me. While there are limits to the goods I can personally achieve, I can retain an interest in their achievement through others for whom I care. Hence, given relations of civic friendliness of a kind Aristotle assumes in an ideal polis, a community in which the largest number of goods can be realized will enrich the lives of all its members. I have little knowledge nor appreciation for opera or the visual arts, but it is better for me that I live amongst those who do, just as it is better for the wen-men that they live in a society where others do appreciate the goods of life above the snow line.

It is in virtue of these point that Aristotle holds that humans can achieve a complete and self-sufficient good only within the polis in which individuals are able to enter a variety of relationships and pursue diverse and distinct goods. The pursuit of these particular goods will be itself a social enterprise that will take place within different associations. The end of the polis is not some completely separate good over and above these partial goods: its end is rather an inclusive end. Hence Aristotle's characterization of the polis in the opening paragraph of the *Politics* as an association of associations.[28] The good that the polis pursues is an inclusive good: it contains all those goods sought in more particular associations. The polis has the comprehensive goal of realizing the good of the 'whole of life'. On this view, the polis does not replace other partial associations, but is rather a community of communities containing a variety of associations realizing particular ends.[29] It has the architectonic function of bringing order to and resolving conflicts between these goods, just as practical reason brings order to the pursuit of the variety of goods pursued by an individual through a life-time. The polis, then, is necessarily plural in the relationships, goods, and associations that it contains. To view the purpose of politics as the pursuit of the good life is compatible with a pluralist view of the political community.

Where the Aristotelian conception differs from certain versions of the modern liberal conception of politics is in the account of the proper relations between different associations should be. On at least some modern versions of liberal neutrality – and I should add I do not think that Rawlsian liberalism belongs to them (section 79 of *A Theory of Justice* contains an excellent articulation of the Aristotelian ideal of social union[30]) – liberalism is founded upon a pluralism of indifference. In our community we do this, in yours you do that and as long as we do not interfere with each other in the pursuit of other ends all is well; I as an individual have this ideal, you have another, and the liberal order allows us to proceed without interfering with each other. The Aristotelian position is based rather on a pluralism of recognition – that a community in which different goods are realized extends the interests of each. Such a pluralism of recognition applies not just to the recognition of the different goods that different communities and associations pursue, but also the different specifications they offer of the good life. In both global and local multi-cultural context one can not only recognize that different communities express relations and attitudes we share at a deeper level in different ways, but find the different modes of expression enriching: the differences capture modes of relating to each other and nature that extend our understanding of such relations. At the same time, pluralism of recognition does not entail that we give up our critical faculties in the face of others and ourselves.[31] The obverse of recognition is criticism. To recognize the virtues of an association or community is at the same time recognize the possibility of vices, say in purely exploitative relationships to both humans and non-human nature.

Narrative

This point about the multiple ways in which a good life can be lived and the plurality of goods it can contain still, however, does not I think capture the extent of variability in which a flourishing life can be led. The Aristotelian position is often associated with an 'objective list' account of the content of wellbeing. A list of goods is offered that correspond to different features of our human needs and powers – personal relations, physical health, autonomy, knowledge of the world, aesthetic experience, accomplishment and achievement, a well-constituted relation with the non-human world, sensual pleasures and so on. Increasing welfare then presented is a question of maximizing the score on different items on the list or at least of meeting some 'satisficing' score on each.[32] This picture of welfare is compatible with the kind of variability I have just outlined. The different items on the list can be satisfied in a variety of different ways, and the descriptions on the list much be such that they allow for different instantiations of the goods. However, while the appeal to such general values is in part right, the maximizing or satisficing approach it suggests is unsatisfactory.

The appeal to an objective list misses the role of history and narrative in appraising how well a person's life goes. The importance of temporal order is already there at the biological level. In appraising the health of some organism, the path of growth and development matters, not just some static scoring system for the capacities it has. Once we consider the cultural and social dimensions of human life this temporal order has a stronger narrative dimension. Answering specific question of how a person's life can be improved is never just one of how one can optimize the score on this or that dimension of the good, but how best to continue the narrative of a life. The question is "given my history, or our history, what is the appropriate trajectory into the future?". This is not to say is there is just one possible trajectory, nor that there is any algorithm to determine it. Our history constrains, it does not determine. However, it moves us away from maximizing concepts and entails a more radical variability. It may be that from some static maximizing perspective, the best course for a person would be to abandon the isolated farm on a fell that he has farmed for the last sixty years and which was farmed by his family before him, and to move to a retirement home. His material, social and cultural life might all improve. But given the way his life has been bound up with that place, that would involve a disruption to the story of his life. The desire to stay 'despite' all the improvements offered is a quite rational one. Likewise, for communities with their particular traditions and histories, what may be 'maximizing' from an atemporal perspective, may not for that particular community.

The significance of narrative is particularly evident in the context of environmental problems and it points to some of the particular problems of the 'wilderness' model of environmentalism discussed in the opening section of the paper. A feature of deliberation about environmental value is that history matters and constrains our decisions as to what kind of future is appropriate.[33] We value natural objects, forests, lakes, mountains and ecosystems specifically for the particular history they embody. That history often includes human use. Most nature conservation problems are concerned with flora and fauna that flourish in particular sites that are the result of a specific history of human pastoral and agricultural activity, not with sites that existed prior to human intervention. This relevance of the past is evident also in the conservation of the embodiments of the work of past generations that are a part of the landscapes of the old world: stone walls, terraces, thingmounts, old irrigation systems and so on. And at the local level past matters in the value we put upon place.[34] The value of specific locations is often a consequence of the way that the life of a community is embodied within it. Historical ties of community have a material dimension in both human and natural landscapes within which a community dwells.

The natural world, landscapes humanized by pastoral and agricultural environments, the built environment all take their value form the specific histories they contain. We enter worlds that are rich with past histories, the narratives of lives and communities from which our own lives take significance. The problem of conservation is, or should be construed as, the problem of *how best to continue the narrative*; and the question we should ask is: what would make the *most appropriate trajectory* from what has gone before?[35] It is because it is about the negotiation of a narrative order between past and future, that it is not simply about 'preserving the past'. Indeed once the significance of the role of a narrative order is understood, we can understand more clearly just what is wrong with that approach. One major problem with the heritage industry is the way it often attempts to freeze historical development. A place then ceases to have a continuing story to tell. The object becomes a mere spectacle, a museum piece, taken outside of any common history. Hence the proper objection from the farmer quoted earlier: "*National Parks, English Nature, they'll finish up with all the farmers running about in smocks, like museum curators.*"

These problems are less visible but still significant in the context of 'wilderness' or 'nature' preservation in national parks of the 'new' worlds. The nature or wilderness to be preserved represents landscapes at a particular point in their history: that of European settlement. The parks of South Africa for example were set up to protect the pristine landscape "*just as the Voortrekkers saw it.*"[36] The comment is echoed in the influential report of the Leopold Committee, *Wildlife Management in the National Parks* in the U.S.A. in which we find the following statements of objectives for parks: "*As a primary goal we would recommend that the biotic associations within each park be maintained, or where necessary recreated, as nearly as possible in the condition that prevailed when the area was first visited by the white man...The goal of managing the national parks and monuments should be to preserve, or where necessary recreate, the ecologic scene as viewed by the first European visitors.*"[37]

The language of wilderness, ecologic scenes, pristine, primitive and natural states, disguises the way in which the history of the landscape is being artificially frozen at a particular point in time To refer to mythical 'natural' or 'wilderness' states avoids the obvious question "why choose that moment to freeze the landscape?". There are obvious answers to that question, but they have more to do with the attempt to forge a national identity than with ecology. In the context of conquest and colonization, the wilderness model of nature has a legitimating role – to render invisible the claims of those whose home one has appropriated.

The good life above and below the snow-line

The Aristotelian position which defends environmental goods as constitutive of the good life is consistent with recognition of the plurality of ways our relations with the natural world can be lived. It is compatible with the recognition of distinct cultural expressions of such relations, the existence of forms of life in which the pursuit of other goods render such relations are marginal, and with the special place the particular histories have in determining what is the appropriate future trajectory for particular individuals and social groups. The good life can be lived both above and below the snow-line. Aristotelian conceptions of public life are quite compatible with pluralism of recognition of the plurality. By the same token they are quite able to provide the basis for robust criticism of the growing authoritarianism of some forms of environmentalism, and in particular the alliance of certain conservation agencies, third world elites and corporate wealth and power which has led to the appropriation and exclusion of indigenous populations in the name of a particular flawed wilderness model of the environment.[38]

Notes

1 From W. H. Auden "Letter to Lord Byron", in: *Collected Longer Poems,* London: Faber, 1968, pp. 59-60.

2 For Aristotle, *"the end and purpose of the polis is the good life"*, in: Aristotle *Politics*, E. Barker (trans.), Oxford: Oxford University Press, 1948, 1280b 38f, where the good life is characterized in terms of the virtues: hence the comment that the best political association is that which enables every man to act virtuously and to live happily. *Ibid.* 1324a, p. 22. In *Ecology, Policy and Politics: Human Well-Being and the Natural World,* London: Routledge, 1993 I argue that a properly constituted relationship is a component of the best life for humans.

3 The argument runs that against the background of modern pluralism, the classical account of politics will necessarily involve the imposition of a particular conception of public virtues, that modern history has shown to be at best authoritarian in its implications, at worst totalitarian. For a good development of this point see C. Larmore: *Patterns of Moral Complexity,* Cambridge: Cambridge University Press, 1987. For a response see J. O'Neill: "Polity, Economy, Neutrality", *Political Studies*, XLIII, 1995, pp. 414-431.

4   Huxley, Aldous: "Wordsworth in the Tropics", in: *Do What You Will*, London: Chatto and Windus, 1929, pp. 113-129.
5   *Ibid.* p.113.
6   *Ibid.* pp.113-4.
7   Anderson, D. and R. Grove: "The Scramble for Eden: Past, Present and Future in African Conservation", in: Anderson and Grove (eds.): *Conservation in Africa*, Cambridge: Cambridge University Press 1987, p. 4. See also R. Grove: *Green Imperialism: Colonial Expansion, Tropical Island Edens and the Origins of Environmentalism 1600-1860*, Cambridge: Cambridge University Press, 1995, and MacKenzie, J.: *The Empire of Nature: Hunting, Conservation and British Imperialism*, Manchester: Manchester University Press, 1988.
8   See G. Monbiot: *No Man's Land*, London: Macmillan, 1994, chs. 4 and 5. Consider, for example the Masai suffering from malnutrition and disease on scrubland bordering the Mkomazi Game Reserve from which they were forcibly evicted from land in 1988 (*The Observer*, April 6th, 1997, p. 12).
9   *The Times*, April 5, 1996.
10 Cited in R. Guha: "The Authoritarian Biologist and the Arrogance of Anti-Humanism", *The Ecologist*, 27, 1, 1997, p. 17. See also the current controversies in Burma.
11 I draw here on M. Walsh, S. Shackley and R. Grove-White: *Fields Apart? What Farmers Think of Nature Conservation in the Yorkshire Dales*, A Report for English Nature and the Yorkshire Dales National Park Authority Lancaster: Centre of the Study of Environmental Change, 1996
12 *Ibid.* p.22. Compare the remarks of farmers on conservation on Pevensey Levels in J. Burgess, J. Clark and C. Harrison: *Valuing Nature: What Lies Behind Responses to Contingent Valuation Surveys?*, London: U.C.L., 1995.
13 *Ibid.* p. 22.
14 *Ibid.* p. 23.
15 *Ibid.* p. 44.
16 O'Neill, John: *Ecology, Policy and Politics: Human Well-Being and the Natural World*, London: Routledge, 1993.
17 For a good discussion of a variety of such positions see F. Arler: "Two Concepts of Justice", *Human Ecology Review*, Vol. 3/1, 1996, pp. 63-76.
18 Nussbaum, M.: "Aristotelian Social Democracy", in: R. B. Douglass, G. Mara and H. Richardson (eds.): *Liberalism and the Good*, London: Routledge 1990, p. 222.
19 Aristotle: *Politics*, E. Barker (trans.), Oxford: Oxford University Press, 1948, 1280b38f.
20 Guha, R. and J. Martinez-Alier: *Varieties of Environmentalism: Essays North and South*, London: Earthscan, 1997.
21 See J. O'Neill: "Thinking Naturally", *Radical Philosophy*, 83, 1997, pp. 36-40.
    The terms 'thick' and 'thin' have become terms of art in debates in political philosophy and ethics. The terms are used, however, to make three different distinctions: 1. Rawls draws a distinction between a thin theory of the good, which refers to the good 'restricted to the essentials' and determines the class of primary goods that rational individuals will necessarily require to pursue whatever ends they might have (Rawls: *A Theory of Justice*, section 60) and a thick theory of the good which specifies particular ends; 2. Williams draws a distinction between thick ethical concepts understood as spe-

cific reason-giving concepts, like brave, cowardly, kind, pitiless, and thin concepts, general abstract ethical concepts, like good, bad, right and wrong (B. Williams: *Ethics and the Limits of Philosophy*, pp. 129-131 and 140ff); 3. Walzer draws a contrast between thick and thin accounts of moral terms, where the thick refers to what is specific to a local place and culture, and thin to what is universal (M. Walzer: *Thick and Thin*, Notre Dame: University of Notre Dame Press, 1994). The three senses are independent. In particular, there is no reason to assume in advance that a thick theory of the good and thick ethical concepts cannot be universally shared but open to local specifications. (See B. Williams: "Truth in Ethics", in: B. Hooker (ed.): *Truth in Ethics*, Oxford: Blackwell, 1996 for a fine, but sceptical statement of this position). That is the position defended in this paper.

22 Aristotle: *Nicomachean Ethics*, 1155a21-22.
23 Herodotus: *Histories*, 3.38.
24 Nussbaum, M.: "Aristotelian Social Democracy", in R. Douglass, G. Mara and H. Richardson, *op.cit.*, p. 217 and passim.
25 Soper, K.: *What is Nature?*, Oxford: Blackwell 1996, p. 214.
26 Aristotle: *Nicomachean Ethics*, 1096b23ff.
27 Aristotle, *op. cit.*, 1166a8.
28 Aristotle: *Politics*, 1252a. 1-7, my emphasis
29 Aristotle: *Nicomachean Ethics*, 1160a8-30.
30 Rawls, J.: *A Theory of Justice*, Oxford: Oxford University Press, 1972, section 79: see the corresponding criticism of the ideal of private society.
31 Compare C. Taylor: *Multiculturalism and 'The Politics of Recognition'*, Princeton N.J.: Princeton University Press, 1992, pp. 61-73.
32 See H. Simon: *Theories of Bounded Rationality, Decision and Organisation*, Amsterdam: North Holland, 1972.
33 The point is developed in more detail in A. Holland and J. O'Neill: "The Ecological Integrity of Nature over Time: Some Problems", *Global Bioethics*, forthcoming and in J. O'Neill: "Time, Narrative and Environmental Politics", in: R. Glottlieb (ed.): *New Perspectives in Environmental Politics*, London: Routledge, 1997.
34 On this see the work of "Common Ground", S. Clifford and A. King (eds.): *Local Distinctiveness: Place, Particularity and Identity*, London: Common Ground, 1993.
35 Holland, A. and K. Rawles: *The Ethics of Conservation*, Report presented to The Countryside Council for Wales. Thingmount Series, No.1. Lancaster University: Department of Philosophy, 1994, p. 37.
36 Reitz quoted in Carruthers, J.: "Creating a National Park, 1910-1926", *Journal of Southern African Studies*, 2, 1989, p. 208.
37 Cited in A. Runte: *National Parks: The American Experience*, 2nd ed., Lincoln: University of Nebraska Press 1987, pp. 198-200.
38 My thanks to Alan Holland and Joan Martinez-Alier for conversations of the issues discussed in this paper.

Biodiversity

The use of 'red lists' as an indicator of biodiversity[1]

Peder Agger and Peter Sandøe

Introduction

The signing of the Convention on Biological Diversity, by 155 countries at the U.N. Conference on the Environment and Development in Rio, 1992, was in some ways an expressive manifestation of the importance attached by the global community to the erosion of biological diversity and the need to take countermeasures. Article 6 of the Convention states that countries shall *"develop national strategies, plans or programs for the conservation and sustainable use of biological diversity."* [2]

The elaboration and implementation of strategies, plans and programs requires among other things species inventories, surveillance, and monitoring activities. Considering the multiplicity of biological forms, spread both geographically and over time, it is obvious that some sort of sampling and selection of indicators is needed if these activities are to be practicable.

In this paper we shall focus on a commonly used indicator, the so-called 'red lists' of threatened species. The basic idea of these lists is that biodiversity is measured by regularly assessing and classifying species and other taxa according their scarcity. Rare, endangered and extinct feature on the list. When the list grows this is a sign of a decrease in biodiversity.

We shall try to clarify and discuss a number of biological and ethical assumptions underlying the use of red lists as a measure of nature protection. The overall aim of the paper is to assess whether red lists really are what they purport to be: A biologically sound measure of biodiversity which relates appropriately to the various ethical concerns for species preservation.

The Red Lists of threatened species

Red lists of threatened species aim to focus attention on vanishing wildlife in the area for which the list is being compiled. They have been one of the most common instruments used internationally in planning and nature conservation, since their introduction by the International Union for Conservation of Nature and Natural Resources (IUCN) in the mid-1960s[3]; and in most nations and many regions, including the globe as a whole, red lists have been prepared. In Denmark even some counties are now compiling and regularly editing red lists, either of all, or some of their wildlife.

The basic idea is that the status of all species, genuses, families and other taxa should be classified and ranked according to a set of internationally accepted criteria indicating the degree of threat. The degrees of threat, or status categories, as developed by the IUCN, can be briefly set out as follows;

Ex: Extinct. Species definitely not localized in the wild over a minimum of 50 years (and in Denmark, since 1850).
E: Endangered. Species in danger of extinction and whose survival is unlikely if presently operating causal factors continue to operate.
V: Vulnerable. Species believed likely to move into the 'Endangered' category in the near future if presently operating causal factors continue to operate.
R: Rare. Species with small populations that are not decreasing at present and not 'E' or 'V' but are always at risk because of their scarcity.[4]

In the Scandinavian countries two more categories are being used:[5]

X: Stands for 'special care'. These are species which, although not endangered, vulnerable, or rare, require special attention because the population has recently decreased drastically – i.e. by more than 50% within the last 25-30 years.
A: Responsibility species: Are species which at any period of time within its life cycle appear in the area concerned in numbers amounting to 20% or more of the world population

The compilation of the lists may vary both in accuracy and difficulty of composition from taxon to taxon and from country to country. In principle the lists are always based on the best available knowledge gathered from literature searches, consultation of local experts and correspondence with specialists from outside. Even so the majority of taxa have usually to be ignored because their status is not known well enough.

In Denmark for example the red list from 1990[6] covers roughly one third of the 30.000 species known as belonging to the native flora and fauna. Of the 9360 species covered 34% are on the list. Of these 353 are in the Ex category and thus have not, as far as we know, bred in the wild on Danish territory since 1850.

In principle the red lists should be revised at regular intervals.[7] In this way the red list, or more precisely the length of it, can be used as an indicator on the general well-being of the flora and fauna.

Problems stemming from lack of data

There are a number of problems connected to the use of red lists. The first and most obvious of these is that it is often difficult to gather the relevant data.

To most non-biologists it comes as a great surprise to learn that only a fraction of the world's animal and plant species have been identified to date. At present 1.7 million species are described and preserved according to international taxonomic standards and thereby recognized by the scientific community. However estimates – more precisely, guesstimates – based on data concerning the rate at which new species are discovered and the thoroughness with which different areas are being researched put the total number of existing species at between 5 and 100 million species. The latest estimate from the United Nations Environmental Programme (UNEP) is 13 million species.[8] Thus according to the most conservative estimate only a third of the earth's species are known.

Red lists do not, then, provide a full picture of species extinction; but in the absence of evidence to the contrary, losses of known species may be pressed into service as indicators of losses of unknown species.

The species records we have are not only incomplete, but biased, because some groups of organisms, like birds and mammals, are more fully recorded than others like viruses, bacteria and fungi.

Hawksworth and Kalin-Arroy, in their comprehensive presentation and discussion concerning the amount and distribution of biodiversity,[9] give these figures for classified organisms:

Bacteria and viruses:	8.000
Protoctista (algae etc.):	80.000
Plantae (fens, mosses, seed plants):	270.000
Fungi (mushrooms etc.):	72.000
Animalia:	1.320.000

Of Animalia the majority, 1.085.000, are Arthropods, 400.000 of which are beetles (Coleoptera).

The current estimate of species classified in Denmark is 30.000. This includes 14 amphibians, 7 reptiles, 185 birds and 50 mammals.

Other factors affecting the data on the basis of which the categorization of species is done are: habitat, the size of the organism, its life cycle and – in the case of animals – its behaviour. Birds, for example, are rather easy to register, whereas registration can be very difficult for certain groups of organisms the visible presence of which typically fluctuates. This is the case for some mushrooms, many orchids and a range of insects which, in one way or another, lie dormant and unnoticed, often underground, for decades, only to burst forth in abundance when conditions become favourable, or after sufficient lapse of time, before returning again to a less conspicuous mode of existence.

Finally, some species are very difficult to identify. For some taxa only a small number of experts are able to make the identification. Even regarding known species the available data are often fragmented and unprecise. It may, for example, be quite difficult to judge whether a species has disappeared for good. This was demonstrated last year by the highly sensational recording in Denmark of a wild population of European pond tortoise (Emys orbicularis) which, unknown to the scientific community, may have survived in Denmark for more than 5000 years.

The underlying notion of biodiversity

Plainly, red lists direct attention to endangered species. More generally they may serve as indicators of biodiversity. An obvious question to ask is whether the lists really do give reliable estimates of biodiversity in the areas where they are applied. However, this question has no single, simple answer, because the very notion of biodiversity is not well defined.

From a biological perspective biodiversity is a relative notion, to be specified with reference to a certain level of organization and with reference to a certain area.

At the genetic level biodiversity may be defined in terms of variation in genotypes. Thus defined the notion of biodiversity may also be applied within a given species. This is, for example, relevant in connection with domestic species, where there is a concern to preserve genetic diversity as a resource for future breeds. However, until now genetic diversity as such has not been focused upon by nature preservationists.

At the species level biodiversity may be defined in terms of number of species found within a certain area. Other levels of organization by reference to which biodiversity may be defined are: population, community, landscape and ecosystem. These higher level entities are, however, more difficult to handle than species. In many cases it has turned out to be impossible to distinguish one wild population of given species from the other. With communities, ecosystems and landscapes it is in pratice even more difficult to decide in a non-arbitrary way where one begins and the other ends.

Red lists clearly record biodiversity at the species level, and in that sense presuppose a particular, elective conception of biodiversity – a conception more fine-grained than, say, landscape diversity, but coarser than genotype diversity. They also generally ignore microorganisms.

These features of the red lists are the result of a number of factors. From a biological point of view the genetic level is the most well defined level. The other levels are inevitably marred by a high degree of vagueness, in some cases (such as with ecosystems) considerable. Some taxa, for example, may contain species and subspecies where there is no clear answer as to where the borders of a species should be set. However, until recently it has only been possible to measure genetic diversity for very few, mainly domestic, species and then only at rather high costs. Therefore from a practical point of view one has to move to the species level to be able to carry out measurements of biodiversity.

There are also considerable practical problems measuring the occurrence of most bacteria, vira and other microorganisms, due to their invisibility and pattern of distribution.

These considerations go some way to explain why measurements of biodiversity are carried out only on plants and animals at species level, as they are in red lists. However, it should not be forgotten that biodiversity is not only a biological notion. It is also a notion used in political and ethical contexts, and there is a wide acceptance that biodiversity per se is valuable and that a loss in biodiversity is bad. The notion of biodiversity thus carries a reference to social and political values prompting nature preservation, and so in this context it may be relevant to discuss and define biodiversity in the light of popular conceptions of nature.

One conception of nature, highly influential in our culture, goes back to the Old Testament. Here nature is viewed as consisting of selection of separate species which can be salvaged from a natural disaster by Noah's Ark. For good reasons microorganisms do not figure in this view.

There is no doubt that the popular and political appeal of red lists at least partly stems from the fact that they fit into this ancient view of nature. Red lists measure how well we have taken care of God's creation. Whether or not people think of themselves as being religious they will often view nature in a way that reflects long accepted religious attitudes. (Also, of course, religious views of nature may be seen as reflecting natural and economic conditions of mankind. Thus it is no wonder that species which have an obvious impact on the life, health and security of the ancient man figure prominently in religious myths.)

However, in many ways the Old Testament conception of nature is at odds with the conception of nature found in modern biology – and the same is probably true of conceptions of nature found in other religions. And this may raise questions about the employment of red lists as a measure of biodiversity.

First, it is now generally recognized that microorganisms play an extremely important role in the household of nature. Secondly, the atomistic view of nature as consisting of a number of separate species has in modern biology been superseded by a more holistic view which focuses on the complex interplay of different kinds of organisms at the level of ecosystems. Also from an aesthetic and recreational point of view macroscopic entities, such as landscapes, may be a much more relevant focus than individual species. Thirdly, we now know that species are not static. Abundance may vary with natural sucession within ecosystems. A species may by natural causes thus become rare and even disappear, and may later on appear again, within a certain area. Species come into existence and may become extinct in the course of natural developments. Thus an enormous number of species became extinct long before human beings came into existence. And the fact that humans have arrived on the scene should not bring the rest of nature to a halt.

Since red lists apparently aim to focus on the human destruction of nature they may be criticized for treating all species as being on a par and not discriminating between those species which become extinct in the course of nature and those for whose disappearance human activities are responsible. In practice, however, the natural extinction is normally considered insignificant, almost nil, compared to the rate of extinction caused by man. And the natural formation of new species, although it certainly goes on, is slow and of no real comfort, given the rate at which species are being annihilated by human activity.

A similar negative bias emerges when we consider the contrast between indigenous animal and plant species, on the one hand, and exotic and domestic species, on the other. The lists only cover species within their natural range. In a country like Denmark where cultural landscapes dominate completely, exotic and domestic species contribute significantly to the local diversity of species. For example in Denmark we have about 2000 wild living vascular plants, out of which 900 are exotic. Although these 900 species contribute significantly to the local diversity they are not included in the red listing.[10]

It soon becomes clear that red lists are not simply used to register biodiversity. To register genetic diversity within a certain area one would also have to count exotic as well as domestic species. The apparent aim of red lists is to register *preservation of indigenous species*.

The notion of an indigenous species is itself a relative notion. Relativity is found in two dimensions. One of these is time. To count as indigenous a species must have been in the relevant area for a considerable period of time. The further back the relevant time limit is set the smaller the number of species which will count as indigenous. The other dimension concerns the question whether the species is present in the designated area 'naturally', independently of human interference, or because human activities, such as agriculture have assisted it. Indigenous species are those which are in the area independently of human interference. This introduces relativity because human activities may affect the living conditions of other species in many ways. The strictest criterion would be to count only those species which would have been in the area if humans never had existed. On this criterion many wild species will not count as indigenous. In a country like Denmark, where the landscape is heavily influenced by agricultural activities going back more than 6000 years, a considerable number of the existing wild animal and plant species will not count as indigenous on this strict criterion.

The decision, as to whether a species should be considered indigenous or not, is left to the judgement of experts. The species will not be considered for inclusion in the category of indigenous species if there is proof, or a strong indication, that it has been deliberately or accidentally introduced.

In one way the practical implications of these observations are limited, for the value of a redlisted species *as indicators* is not affected greatly by which species are selected, assuming the number is suffiently large. But to the extent that red lists are used as measures of nature conservation the confinement of redlisting to what is indigenous to a region will reflect underlying values. While the claim that the preservation of indigenous species is an adequate measure of nature protection to some extent is in harmony with generally accepted views concerning nature conservation, and most people will regret the loss of indigenous species and will favour initiatives to preserve such species, it may well be argued that there is much more involved in nature preservation than the protection of indigenous species. To many people the protection of non-indigenous species is also a matter of great concern. Consider for example the protection of traditional meadows with their characteristic vegetation, or the conservation of old and local breads of domestic animals. Both are examples of measures to ensure the continuing presence of largely non-indigenous species – exotic biodiversity if you will.

Thus again it turns out that underlying the use of red lists as measure for nature conservation is a narrow conception of biodiversity.

Conclusion

As has become clear from the discussion in the previous section biodiversity is a multidimensional concept, a concept in need of careful handling and about whose precise definition, in a given context, we should always seek to be clear. The dimensions of biodiversity may be summarized in the following way:

Dimensions of biodiversity

1) *Geographic scale*
   Focus may be on areas of very different sizes
2) *Level of organization*
   Focus may be on diversity of genotypes, species, landscapes, ecosystems etc.
3) *Size of organism*
   Focus may be on diversity of macroorganisms like vertebrates and vascular plants, or on microorganisms like bacteria and vira.
4) *Cause of extinction*
   Focus may be on all losses of species or only on those losses which are caused by certain human activities
5) *Age*
   Focus may vary according to how long the species (or other relevant entity) has been in the relevant area. One extreme is counting only species which have been around for a very long time
6) *Degree of independence of human interference*
   Focus may be on indigenous species only, or may be broadened to include domestic and exotic species also

In connection with all these dimensions red lists will take a certain stance. We have argued that in many cases the stance taken is tendentious and, where both ethical and biological issues are connected, of debatable suitability.

The main answer to these objections will probably be that we have nothing better. According to IUCN the red lists provide an easily and widely understood method for highlighting species under extinction risk. And the lists are used. The snail darter (a rare species of fish) and spotted owl issues in US have for various reasons been notorious. But many other instances of red list use demonstrate that the lists play a significant role in contemporary nature conservation. This is also the case in Denmark where e.g. the Minister of Environment and Energy recently confirmed that the red list is the basis for national species policy.[11] The IUCN categories have been widely recognized internationally for years, and are now found in a whole range of publications and listings all over the world which year by year add to the reference value of the red list system.

But the IUCN also wishes to revise the categories and new definitions are in process. The general aim of the new system is to provide an explicit, objective framework for the classification of species according to their extinction risk.[12] Some new, more specific categories have been added (see Fig.) and the criteria for each of them have been made more exact by introducing percentile changes over a defined number of generations or years, and bringing in areas and absolute numbers of mature individuals.

Our main message is that in the process of revising and refining the use of red lists it is important to discuss the biological and ethical issues raised above. Only in this way can red lists become a biologically sound means of management of a biodiversity which relates appropriately to widely felt ethical concern about the protection of the natural world.

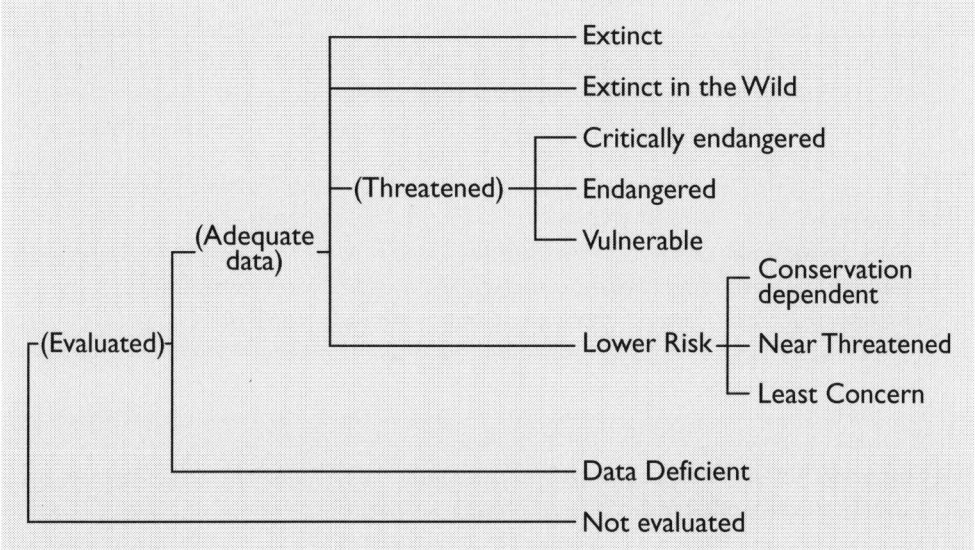

Text to figure:
In the coming red list from IUCN the classification will be more detailed. The old category 'Extinct' and 'endangered' are split, and three new categories are introduced for species that formerly would not have been listed: 'Conservation Dependant'; 'Near Threatened'; and 'Least Concern' (source: see note 12).

Notes

1 Thanks are due to Finn Arler, Jørn Jensen and Ingeborg Svennevig for useful comments to earlier versions of this paper. Special thanks go to Paul Robinson for clarifying thoughts and improving the English language throughout the paper.
2 Article 6 (a) in The Convention on Biological Diversity, 5 June 1992. Here cited from Prip, C., P. Wind, and H. Jørgensen (eds): *Biological Diversity in Denmark – status and Strategy*, Copenhagen: Ministry of Environment and Energy 1996, p. 7
3 IUCN: *1986 IUCN Red List of Threatened Animals*, Cambridge: The IUCN Conservation Monitoring Centre, 1986.
4 Already in 1986 IUCN also used some more categories: 'I' for indeterminate, 'K' for insufficiently known, 'T' for threatened, and 'CT' for commercially threatened. To day a more detailed and objective set of criteria are being introduced. A new category is 'EW' for extinct in the wild e.g. the Californian condor which survives only in captivity. See also note 12.
5 Nordisk Ministerråd: *Hotade djur och växter i Norden*, Stockholm: Nordisk Ministerråd, 1978.
6 Asbirk, S. og S. Søgaard (eds.): *"Rødliste 90" – særligt beskyttelseskrævende planter og dyr i Danmark*, Copenhagen: Miljøministeriet, Skov- og Naturstyrelsen, 1991.
7 The Danish 1995/96 issue is expected to be published spring 1998.
8 UNEP (United Nations Environmental Programme): *Global Environment Outlook*-1.- http:77www.grida.no/geo1/exsum/ex3.htm. Nairobi January 1997.
9 Hawksworth, D.L. and M.T. Kalin-Arroyo (eds.): Magnitude and Distribution of Biodiversity, in Heywood, V.H. and R.T. Watson (eds.): *Global Biodiversity Assessment*, Cambridge: UNEP & Cambridge Univ.press 1995, pp. 107-173.
10 Vind, P.: Rødliste 95. *URT*, 1997. 21:2, pp. 3-9.
11 Miljø- og energiministerens besvarelse af spørgsmål nr.1888 stillet af Folketingsmedlem Jette Gottlieb (EL) 18. april 1996. (Answer to a question in the Parliament to the Minister of Environment and Energy, no.1888, 18. april 1996).
12 IUCN: IUCN Red List Categories prepared by the IUCN Species Survival Commission, As approved by the 40th Meeting of the IUCN Council. Gland Switzerland. 30 November 1994. pp. 21. New categories will be added. 'CR' for critically endangered; 'LR' for lower risk which can be separated into three subcategories: 'cd' for conservation dependant, 'nt' for near threatened and 'lc' for least concern. 'DD' for data deficient and 'NE' for not evaluated.

Nature, culture and biodiversity[1]

Kay Milton

> "... *the thing about beaches that makes them so important to me is their naturalness. They are places that literally have a life of their own, where rhythms of tides and seasons set an agenda that seems to stand outside human time.*"[2]

Introduction

One of the main ways in which social science contributes to social movements is through constructive critique. By analyzing key concepts and exposing contradictions, social scientists can help to improve the credibility of arguments. Movements such as socialism and feminism advanced through this kind of treatment, and so has environmentalism developed through the scrutiny of scholars such as Naess, Dobson and Goodin[3] among many others. Social anthropologists are well-qualified to play this role,[4] since one of their central concerns is the relationship between thought and action, how people's activities shape and are shaped by their understanding of the world. This relationship is fundamental to the process of social reform, since the credibility of advocacy can depend on the consistency of thought and action; whether the actions advocated by a social movement are in accordance with its world view.

There is a counter argument to this, however. Bourdieu pointed out that "*practice has a logic which is not that of logic*" and that this has to be acknowledged, "*if one is to avoid asking of it more logic than it can give*".[5] In other words, the consistency sought after by analysis is not necessarily appropriate for practical activities. Perhaps, for their own practical purposes, and for trying to influence the practices of others, social reformers are better off with muddled arguments. If so, then we can advance our understanding of the relationship between thought and action by asking why.

Environmentalists are social reformers. They try to persuade people, at least in western and industrial societies, to act differently than they have in the past, to replace harmful ways of using their environment with more benign ones. As they do so, they express and expose, both implicitly and explicitly, the assumptions, values and objectives that guide their advocacy. The effort to conserve biodiversity has become a major component of contemporary environmentalism. It was the subject of the *United Nations Convention on Biological Diversity* (the Biodiversity Convention), a product of the 1992 Rio Earth Summit. It is the central concern of some of the major international non-governmental organizations (NGOs), such as the World Wide Fund for Nature (WWF) and the World Conservation Union (IUCN), and is treated as an important indicator of sustainability.[6]

In this paper I examine the significance for biodiversity conservation of a fundamental opposition between natural processes and human activities, often expressed by anthropologists as a dichotomy between nature and culture. Anthropologists have been interested for some years in whether this dichotomy is universal in human thought or characteristic only of some societies. They have not reached a consensus on this issue, but appear to agree that it is fundamental in western thought, in which nature and culture are often presented as opposed and in conflict.[7] The nature – culture dichotomy enters the conservationist world view in two ways: it acts as a source of values for conservationists, and it is present in the scientific knowledge which many of them treat as a factual basis for their activities.

The analysis that follows examines the logic of conservationist thought. I shall argue that the nature – culture dichotomy, when combined with a dependence on science, creates logical inconsistencies in the conservation world view. In section two, I examine the role and character of the dichotomy in conservationist thought and show how difficulties are created by its combination with science. Section three, by focusing on the concept of species, shows how these difficulties are also present in the scientific knowledge on which conservationists draw. The final section considers what consequences might follow if conservationists were to abandon the nature – culture dichotomy, and ends with a brief consideration of why they retain it.

I should make clear that the conservationist world view presented here is a very specific one, derived from observations of statutory bodies and NGOs in Britain and Ireland,[8] and from documents produced by these and international institutions such as the United Nations and the European Union. The understanding of conservation as a science-based activity comes from these organizations.[9] I am aware that conservation discourse in different western societies, and in different sectors of the same society, is characterized by a variety of concerns and emphases; for instance, religious, political and economic arguments are often used alongside or instead of science to substantiate a conservation ethic. Readers can judge for themselves how far the analysis presented here is generally applicable.

Nature and culture in the conservationist perspective

The conservation of biodiversity is a relatively new way of describing what has been an established public concern for many years in some western countries. 'Nature conservation', which encompasses concern for biodiversity, has long been institutionalized, particularly in Britain and America, in the activities of NGOs and statutory agencies. It is easy enough to locate a concept of nature in the nature conservation world view. 'Nature' is what conservationists strive to conserve, and if they did not distinguish it somehow from what is not nature, they would strive to conserve everything, which, clearly, they do not. They continually use labels which indicate that human products are qualitatively different from 'natural' products. 'Primary forest', produced by nature without human interference, is distinguished from 'secondary forest', which grows after land has been cleared for human use. 'Pristine' habitats are distinguished from those affected by human activity. Native species are distinguished from those which people have introduced, and wild varieties are distinguished from those created through biotechnology. In general (though not exclusively), it is the primary and pristine habitats, the native and wild species, about which conservationists are concerned.[10] I suggest that this constitutes a version of the nature – culture dichotomy, as anthropologists have described it, present in the minds of conservationists and used by them to inform their actions and decisions.

Nature, culture and value

An important characteristic of the conservation perspective is the differential value accorded to natural and cultural products. Primary forest is not only different from secondary forest, it is considered, in principle, more desirable. Pristine peatlands are valued more highly than areas that have been cut over and regenerated, and native species are generally considered more important than those introduced, deliberately or otherwise, as a consequence of human activity. 'Naturalness', seen in terms of the absence of human influence, is one of the criteria used to select nature reserves and other protected sites.[11]

The differential value placed on natural and cultural things creates a dilemma for conservationists which they openly acknowledge. Conservation itself is a human activity, one which is dedicated, in many instances, to the deliberate alteration of habitats, and in some cases to the deliberate introduction of species. Conservationists often acknowledge that some of the most valued ecosystems are products of human activity. Ancient hedgerows, originally planted as field boundaries, are considered vital for the conservation of farmland wildlife.

Gravel pits can become important habitat for wildfowl and abandoned quarries are often significant botanical sites. So some cultural activities, it seems, *are* valuable. The important criterion is that these activities are seen as benefiting nature. An agricultural landscape is important if it serves the needs of non-human organisms; a building is important if it houses a colony of bats. So when conservationists campaign to save these cultural products, they do so to help nature. Species deliberately introduced into an area by conservationists are those which were once present but have become extinct as a result of human activity. Conservationists would prefer not to have to do these things, but nature is seen, in many respects, as fragile in the face of cultural forces; it has to be protected and, where necessary, restored or recreated.[12]

Biodiversity

The most important characteristic of nature, in conservationist thought, is its diversity. It is valued partly for reasons of human interest, because it is seen as a source of material and spiritual benefit to human beings. But, particularly for conservationists who base their arguments on science, diversity is seen as serving the needs of life itself. The greatest possible diversity of living things is seen as benefiting the process of evolution, to which all life forms are assumed by many[13] to owe their existence. The more diversity there is, the greater is the chance that some life forms will be able to adapt to changing conditions, and that life itself will continue.

Conservationists' concerns for natural diversity have been given official sanction by the international discourse on biodiversity (biological diversity). According to the Biodiversity Convention, biodiversity includes "*diversity within species, between species and of ecosystems.*"[14] It is often expressed more simply as "*the variety of life.*"[15] In the UK, the effect of the Biodiversity Convention was to bring government obligations closer to some of the long-standing objectives of the conservation NGOs, and to create what is widely acknowledged as a constructive partnership between government and the voluntary sector, in the effort to meet the requirements of the Convention.[16]

The emphasis on biodiversity has important implications for the conservationist perspective. Most obviously, it requires whatever is rare to be valued more highly than what is common. The fact that conservationists have always placed more value on rare species and ecosystems indicates the extent of their concern for diversity in nature, rather than simply for nature itself. The aim of conservation is to produce conditions which maximize diversity. So, for instance, the populations of common species which threaten the survival of rarer species are controlled to give the rarer species the chance to increase. Population control is considered particularly appropriate for commoner species which are not indigenous (such as the grey squirrel and the ruddy duck in Europe)[17] and for those which have become common as a result of human activity (such as gulls and crows in Britain and Ireland).

Thus far, I have presented the conservation world view as consisting of several components which articulate coherently as follows: nature and culture are separate and culture can both harm and benefit nature. Because nature is more valuable than culture, those cultural activities that benefit nature are desirable and those that harm it are undesirable. Nature's most valuable asset is its diversity which, in the long term, offers it the best chance of surviving the harmful effects of culture. The conservation objective is to help nature achieve this goal by maximizing biodiversity. I shall argue that the coherence of this perspective breaks down once another ingredient, scientific knowledge, is taken into account.

The reliance on science

There are many different ways of describing the diversity of nature. J. Maclair Boraston's book, *British Birds and their Eggs*,[18] contains chapters on black birds, black and grey birds, brown birds with spotted breasts, long-legged birds, and so on, and ends with a chapter on *"birds which do not fall into groups"*. The book was written for the beginner; it presents birds in terms of their most obvious characteristics, enabling the reader to locate their descriptions without any knowledge of their scientific classification. This is a simple illustration of what conservationists mean by biodiversity. Beyond the obvious differences among living things are many subtle variations which cannot be detected by an inexperienced observer, or even by an expert without sophisticated equipment. All this diversity is described by science in terms of species, subspecies and higher order classifications (genera, families, orders, and so on). For those scientists who accept the theory of evolution, these differences are produced by a combination of mutation, selective environmental pressures and reproductive isolation.

In the conservationist perspective described here, biodiversity conservation depends on science. Scientific knowledge defines biodiversity and supplies the criteria and the technology for measuring it. It also provides the basis for deciding how to conserve biodiversity. The scientific discipline of ecology reveals what organisms need for their survival, how they interact with others of their own and different species, what factors affect their health and numbers. The scientific understanding of these relationships has given rise to the view that ecosystems are distinct repositories of biodiversity, making it important, in the conservationist perspective, to protect whole ecosystems, as well as species and populations.

In relation to other features of the conservationist world view, this dependence on scientific knowledge looks a little strange, for science is, after all, a human activity. It therefore falls on the culture side of the nature – culture divide, and to give it overriding importance, to allow it to define natural categories and set the goals of conservation, seems inconsistent with the greater value conservationists attach to nature over culture. But science, in the conservation perspective, is no ordinary cultural activity. It may belong to culture, but it provides a window onto nature; it teaches us what nature is 'really' like. Scientific knowledge is treated by conservationists as revealed truth.

It is the claimed truth of science that makes the conservation perspective inconsistent, for if science generates knowledge of what the world is 'really' like, then the fundamental division between nature and culture becomes untenable. Evolutionary biology, on which many conservationists draw for their understanding of biodiversity, locates humanity within nature, as a product of the same forces that shaped other species. In this context, to suggest that human activities lie outside nature makes no more sense than to suggest that what lions or bumble bees do lies outside nature. To suggest that a farmed landscape or a city, is unnatural makes no more sense than to say that a coral reef or a termite mound is unnatural. Thus, while providing a factual basis for the conservation of biodiversity, scientific knowledge invalidates the model from which that same enterprise derives its values, the nature-culture divide. But the inconsistency goes deeper than a clash between facts and values, for it is present in science itself. This argument can be developed by focusing on one particular dimension of biodiversity, the distinctiveness of species.

Keeping species apart

Biodiversity, as we have seen, is defined in terms of three dimensions: variation within species, between species and between ecosystems.[19] The effort to conserve biodiversity takes place on all three levels, but that which has the highest public profile is undoubtedly the effort to conserve species. In most instances, this means preventing a species from becoming extinct by protecting the lives of its remaining members (for instance, by conserving their habitat or controlling predation). But extinction is also understood in another way: a species can become extinct by losing its distinctiveness. This happens when species which are closely related to each other interbreed and produce fertile offspring which in turn reproduce, either amongst themselves or with members of the two parent species. The result is a dilution of the gene pool; extinction in this form is not death, but loss of genetic purity. There are many examples: Australian wild dingoes interbreeding with feral dogs, Scottish wild cats interbreeding with domestic cats and white-headed ducks in Spain interbreeding with the non-native ruddy ducks which have escaped from wildfowl collections and spread across Europe.

On the face of it, the recognition that species can interbreed appears to contradict one of the main ways in which the concept of species is understood by biologists. Mayr defined species as *"groups of actually or potentially interbreeding natural populations which are reproductively isolated from other such groups."*[20] Conservationists treat white-headed ducks and ruddy ducks as separate species,[21] and yet in Spain, where they are no longer reproductively isolated from each other, they constitute an 'actually interbreeding population'. Some biologists might argue, therefore, that they constitute one species, not two. However, this argument overlooks Mayr's qualification that species are *natural* populations. Put another way, *"a species is a population whose members are able to interbreed freely under natural conditions."*[22] If interbreeding between populations takes place only under non-natural conditions, those populations do not, in Mayr's view, constitute a single species.

Those biologists who distinguish between natural and unnatural conditions generally do so, as conservationists do, in terms of human influence. The presence of feral dogs in Australia, domestic cats in Scotland and ruddy ducks in Europe are all the result of human activity. Thus the view that ruddy ducks and white-headed ducks, for example, are distinct species is sustained by excluding human activities from the sphere of nature. But, as I have argued, evolutionary biology will not permit this, since it treats human beings as a species in the same way as any other. Thus the nature – culture dichotomy makes at least some of the knowledge generated by science logically unsound.[23] In doing so it confounds, not only the values employed in conservation, but also what conservationists see as the factual basis of their activities.

The obvious way of correcting this deficiency is for biologists and conservationists to abandon the nature – culture dichotomy. The case for doing so can be developed with reference to a recurring subject of debate in biology and philosophy: the question of whether species are real natural categories or cultural artefacts.

Are species 'real'?

It is important, for the purpose of conservation, that species exist as entities in nature and not just in the minds of scientists. Only if species are real natural categories can conservationists who strive to protect them legitimately claim to be conserving *nature*. Some biologists acknowledge that some of the categories they recognize belong to culture rather than nature. Wilson pointed out that the higher order categories – genus, family, order, class and so on – "*are a mental construct invented for convenience,*" and that their exact limits are arbitrary.[24] They help us to make sense of nature, but play no active part *in* nature; they describe the products of natural processes but they do not affect those processes. Species are viewed rather differently; biologists generally assert that they are real natural categories, "*in some way objective or existing independently of the classifier.*"[25]

In order to address this issue it is important to identify two separate claims implied here. It is one thing to claim that, in nature, *there are such things as species*. It is quite a different matter to claim that the species recognized by biologists are those that exist in nature. When biologists argue that species are real, they are usually making the first of these claims: that species exist in nature. The question of whether their allocation of organisms to species is accurate has to remain open in order for knowledge of biodiversity to advance. Biologists are constantly reviewing and refining their classifications in the light of new discoveries.

What convinces biologists that species are real, in the sense that they play a part *in* nature, is the manner in which their defining criteria coincide. Mayr's understanding of species as reproductively isolated populations is just one of several commonly used definitions. Species are also defined in terms of morphological and genetic similarity. These criteria coincide in the sense that organisms which are genetically similar tend also to be morphologically similar, and organisms which can interbreed freely tend to resemble each other genetically and morphologically.[26]

However, in terms of two of these criteria, the boundaries between species are no less arbitrary than the boundaries of the higher order categories. In order to identify species in terms of morphological or genetic similarity, biologists need to decide how much difference is necessary in order to distinguish one species from another. They do not do this consistently, at least, not in a way that coincides consistently with the ability to interbreed. Some organisms which can interbreed successfully are less similar to each other, in terms of morphology and genetics, than others which do not.[27] On the other hand, the ability to breed successfully is a non-arbitrary indicator of a natural category. The production of fertile offspring is a sign of something going on in nature, and not just in the minds of analysts. Thus, biologists could strengthen their case that species are real natural categories if they defined them solely in terms of the ability to breed successfully.[28] This is not to suggest that genetic and morphological similarity are not important characteristics of species, only that their variability makes them unsuitable as defining criteria.

The importance of breeding success as a criterion for identifying species is that it allows organisms to define their own categories, and that this confirms their significance in the world beyond human understanding. However, the nature – culture dichotomy undermines the value of this criterion by creating a disjuncture between observed reality and biologists' interpretations. It is clear that, in some cases, individuals belonging to what some biologists regard as different species are capable of interbreeding when brought together as a result of human activity. The processes taking place are essentially the same as those that occur when members of the same species reproduce, and yet the nature – culture dichotomy leads biologists to interpret them differently. In such instances, the organisms are not being allowed to define their own categories, but instead have biologists' categories thrust upon them.

Conservation without nature and culture

How would our understanding of biodiversity be affected by the loss of the nature – culture dichotomy? Before addressing this question, it is important to clarify what it implies. Abandoning the dichotomy means accepting the natural status of human beings, treating human activities and their products as part of the natural order. This does not mean giving up our concept of nature; it means defining nature as the entire scheme of things to which human beings and everything else belong. Nor does it prevent us from distinguishing culture from nature, only from treating it as separate from and opposed to nature. Instead, culture would be seen as a part of nature. This accords with Goodin's "green theory of value", which states that people value nature because it gives them a context in which to set their own lives, that human projects acquire meaning by being a part of something larger.[29] In other words, nature is important precisely because it encompasses human processes and products.

The understanding that humanity is a part of nature plays a prominent role in contemporary environmental discourse, and is fundamental to what are often described as 'ecocentric' viewpoints. It is also acknowledged by organizations which adopt the perspective analyzed above,[30] but as long as the terms 'natural' and 'non-natural' are used to describe non-human and human processes respectively, this acknowledgement amounts to nothing more than lip service. Once human activities and their products are treated unambiguously as part of nature, the distinction between natural and non-natural, and the classifications that depend on it, disappear.

The effect of this on our understanding of biodiversity need not be radical. Populations which interbreed freely only as a result of human intervention, such as the ruddy duck and white-headed duck, the Scottish wild cat and the domestic cat, would be treated as single species. The diversity of these populations would not, of course, be affected by our reclassification of them; we would simply describe it as *intra*-species rather than *inter*-species diversity. Since variation within as well as between species is already valued as an aspect of biodiversity, conservationists could go on campaigning to save the white-headed duck and the Scottish wild cat on the grounds that they represent intra-species variation.[31]

In addition, abandoning the nature – culture dichotomy would simplify the evaluation of ecosystems as repositories of biodiversity. At present, conservationists assess the value of ecosystems partly in terms of their naturalness, their freedom from human influence; the convention imposed by the nature – culture opposition requires them to do this. But, as we have seen, this criterion is not consistently applied. Products of human activity are valued if they benefit nature by supporting biodiversity. By abandoning the nature – culture dichotomy, and with it the distinction between natural and non-natural processes, conservationists would leave themselves free to focus on conserving biodiversity wherever and however it occurs.

Conservationists commonly raise two objections to the suggestion that human beings and their activities should be treated as part of nature. First, they point out that this would place organisms created through biotechnology on a par with wild organisms. This is so, but if the purpose of conserving diversity is to ensure the continuation of life through evolution, then it does not matter whether that diversity is the product of human or of non-human processes. What *does* matter is that human activity should not reduce diversity, that it should not result in the replacement of many varieties with few varieties. In this context, the conservation of both wildlife and of 'rare breeds' (i.e. rare domestic and cultivated varieties) are important.

Second, they point out that treating human activities as natural could be taken as justification for ecological destruction. In other words, if it is natural for human beings to chop down the rainforests and pollute the seas, is it not right that they should do so?[32] But this point is only valid if naturalness is seen as conferring value. Once we abandon our grounds for distinguishing natural from non-natural processes, it becomes nonsensical to value something in terms of naturalness. In addition, if biodiversity *per se* were treated as the guiding value, then all activities which reduce biodiversity would be undesirable.

But would it work?

If, as I have argued, abandoning the nature – culture dichotomy would make the conservation perspective more coherent and simpler to apply, it is worth asking why conservationists retain it. This question is worth far more attention than I can give it here, but I suggest that part of the answer lies in the social context in which conservationists operate. They are, as I have said, social reformers. They need to get a message across, to communicate with those who do not share their views and persuade them to change their ways. In their advocacy they need to be governed more by what works than by logical coherence.

In western industrial societies, as many analysts have pointed out, the nature – culture dichotomy is a familiar model. By using it, environmental campaigners speak to their audience in terms which they know will be understood, and they have found the perceived 'otherness' of nature a useful starting point for eliciting a sense of guilt and responsibility towards non-human species. As Adams expressed it, *"... conservation needs to be built on a foundation of individual awareness of and concern for nature ... it is here, on the ground of the relations between people and nature, that conservation must stand if it is to move forwards."*[33] If they abandoned the nature – culture dichotomy, conservationists might lose an idiom vital to the effective communication of their message, something for which it is worth sacrificing a degree of logical consistency.

Notes

1. I am grateful to John Stewart, Peter Bowler, Iwan Morus, Ingeborg Svennevig, Finn Arler and participants in the seminar on the Cross-cultural protection of nature and the environment, at the Humanities Research Center, Odense University, for their comments on earlier drafts of this paper.
2. Adams, W. M.: *Future Nature: a vision for conservation*, London: Earthscan 1996, p. 3.
3. Naess, A.: *Ecology, Community and Lifestyle*, Cambridge: Cambridge University Press 1989. Dobson, A.: *Green Political Thought*, London: HarperCollins 1990. Goodin, R. E.: *Green Political Theory*, Cambridge: Polity Press 1992.
4. See Milton, K.: 'Introduction: Environmentalism and Anthropology', in: K. Milton (ed.): *Environmentalism: The view from anthropology*, London and New York: Routledge 1993.
5. Bourdieu, P.: *Outline of a Theory of Practice*, Cambridge: Cambridge University Press 1977, p. 109.

6   See *Sustainable Development: the importance of biodiversity*, Sandy, Bedfordshire: Royal Society for the Protection of Birds 1996.
7   I would argue that the dominance of an opposition between nature and culture in western thought is often exaggerated, but this is not my concern here (see Milton, K.: 'Nature and the environment in indigenous and traditional cultures', in: D. Cooper and J. Palmer (eds): *Spirit of the Environment*, London and New York: Routledge, in press).
8   In particular, I draw on the work of organizations such as the Royal Society for the Protection of Birds (RSPB), the Wildlife Trusts, WWF(UK), English Nature (and its predecessor, the Nature Conservancy Council) and on publications produced by these and other bodies in Britain and Ireland.
9   In the UK, the most commonly used statutory mechanism for nature conservation is the Site/Area of Special Scientific Interest (SSSI or ASSI); its equivalent in the Republic of Ireland is the Area of Scientific Interest (ASI, see note 11).
10 Further discussion of these issues can be found in *Nature Conservation in Great Britain*, London: Nature Conservancy Council 1984, and Nicholson, M.: *The New Environmental Age*, Cambridge: Cambridge University Press 1987.
11 See Ratcliffe, D. A. (ed.): *A Nature Conservation Review*, Cambridge: Cambridge University Press 1977; *Areas of Scientific Interest in Ireland*, Dublin: An Foras Forbatha 1981; Moore, N.: *The Bird of Time*, Cambridge: Cambridge University Press 1987.
12 See Milton, K.: *Environmentalism and Cultural Theory*, London: Routledge 1996, p. 124.
13 Of course, not all scientists are evolutionists. Some attribute biodiversity to other sources (such as divine creation). Here I am concerned specifically with scientific knowledge in which biodiversity *is* attributed to evolution, a view which appears to predominate in the western discourse on biodiversity.
14 *Biodiversity: The UK Action Plan*, London: Her Majesty's Stationery Office 1994, p. 6.
15 *Biodiversity: The UK Steering Group Report*, London: Her Majesty's Stationery Office 1995, Vol. 1, p. 12.
16 *Biodiversity: The UK Steering Group Report* (see note 15) is a product of this partnership.
17 See Joint Nature Conservation Committee: *Annual Report 1993-94*, Peterborough: JNCC 1994, p. 17-18.
18 Boraston, J. M.: *British Birds and their Eggs, with a simple method of identification*, London: W. & R. Chambers Ltd 1908.
19 See *Biodiversity: The UK Action Plan*, London: Her Majesty's Stationery Office 1994, p. 6 and *Biodiversity Challenge: an agenda for conservation in the UK*, Sandy, Bedfordshire: The Royal Society for the Protection of Birds *et al.* 1993, p. 1.
20 Mayr, E.: *Systematics and the Origin of Species*, New York: Columbia University Press 1942, p. 121.
21 Lawson, T.: 'Brent duck', in *Ecos* 17 (2) 1996.
22 Wilson, E. O.: *The Diversity of Life*, London: Penguin Books 1992, p. 36, emphasized in original.
23 Cf. Ellen, R.: 'Introduction', in R. Ellen and K. Fukui (eds): *Redefining Nature*, Oxford: Berg, p. 10.
24 Wilson, E. O.: *The Diversity of Life*, London: Penguin Books 1992, p. 145.
25 Ruse, M.: *The Darwinian Paradigm: Essays on its History, Philosophy and Religious Implications*, London and New York: Routledge 1989, p. 97.

26 Ruse, M.: *The Darwinian Paradigm: Essays on its History, Philosophy and Religious Implications*, London and New York: Routledge 1989, p. 113.
27 See Jones, S. quoted in Lawson, T.: 'Brent duck', in *Ecos* 17 (2) 1996, p. 30.
28 I acknowledge that this can apply only to organisms that reproduce sexually. The definition of species for asexually reproducing organisms is more problematic (see Wilson, E. O.: *The Diversity of Life*, London: Penguin Books 1992). In addition, I am not intending to imply that species will always be clearly distinguishable empirically, in terms of the ability to breed successfully. Speciation, as described by the theory of evolution, is a gradual process. At any one time, the status of some populations as separate species is bound to be unclear.
29 Goodin, R. E.: *Green Political Theory*, Cambridge: Polity Press 1992, p. 30ff.
30 See, for instance, *Nature Conservation in Great Britain*, London: Nature Conservancy Council 1984, p. 7.
31 It is worth pointing out that, according to one version of evolution theory, preventing hybridization may reduce rather than enhance biodiversity. In terms of Dawkins' 'selfish gene' theory (Dawkins, R.: *The Selfish Gene*, Oxford: Oxford University Press 1976), genes become endangered when they are isolated in small populations. The best chance of survival for a gene stuck in an endangered population may be hybridization with a more common population. Biodiversity is maintained because genes that would otherwise have died out live on in the hybrids.
32 Lawson, T.: 'Brent duck', in *Ecos* 17 (2) 1996, p. 31.
33 Adams, W. M.: *Future Nature: a vision for conservation*, London: Earthscan 1996, p. 9, emphasis added.

Increase of biodiversity through biotechnology: Genetic pollution or second order evolution?

Merete Sørensen

Introduction

Our biosphere is a manifestation of evolution. Species gradually change, and new species occur. Nevertheless, some of us find it defensible to ascribe *authenticity* and *integrity* to nature; and some are worried even, that the authenticity and integrity of nature is disturbed (not to say spoiled) by human interference. According to some, nature already has been subject to a so-called *genetic pollution* as a consequence of human cultivation of plants and selective breeding of animals, whereas others ascribe genetically polluting effects only to the modifications resulting from applications of the kind of genetic engineering, which transcends species boundaries.

Both parties reject the claim, that we can obtain an increase of biological diversity through intentional construction of new organisms by biotechnologies, and see it as a perversion. They will in no way accept, that new races or species resulting from technological intervention are categorized at the same level as those which have evolved in a more natural way. In particular, they resist the implicit possibility that the humanly constructed types of organisms could substitute extinct species, i.e. weigh up the *decrease* of biodiversity caused by other human activities. It is exactly this possible implication, which makes it urgent for them to argue for the incommensurability of the processes of evolution on the one hand, and the human intentional modifications of organisms on the other. They emphasize the existence of an *essential* difference between the slow, sound evolutionary growth of biodiversity and the rapid, technology aided increase of the number of species. According to some of them this difference can best be compared with the division processes in a healthy cell and a cancer tumour, respectively.

On the other hand, there are people who argue that the positive connotations attached to the organisms, which are minimally influenced by human activities, are displaced. Humans and their technical skills are themselves products of evolution. Accordingly, the use of these skills can most adequately be seen as means for evolution to accelerate its own course, i.e., as a *second order evolution*. In this perspective, the protection of the genomes of not intentionally changed organisms might rather represent a stagnation of evolution, an obstacle to the full development of its potentials. In general, there is no reason to fear the use of our skills to change the genomes of organisms by virtue of genetic engineering. The wish to preserve the natural evolution of non-human organisms from human interference is regarded as a romantic dream, ignorant to the fact that the actual mutations even of wildlife are increasingly influenced by chemicals, pesticides, radiation and similar consequences of human activity. The relevant alternative, therefore, is not: human impact or no human impact on the non-human part of evolution, but: more or less human impact. Consequently, there is no sharp distinction between a "natural" and a technology aided increase (or decline) of biodiversity.

I shall examine the content of these two positions, which seem to be so charged with opposite values, by asking which one is the more fruitful perspective. My focus will be on *genetic engineering*, and for the sake of simplicity I have limited the discussion to *animals*. I see it as one of the most urgent tasks to discuss this particular area, partly because genetic engineering is of growing importance, and partly because it is not possible to deal with central ethical aspects of genetic modifications of animals within the at present most dominant theory of animal ethics: welfare ethics. Animal welfare is a very important aspect of animal ethics, of course, but from my point of view, the theory does not catch the essence of the ethical perspectives, which are opened up by the most advanced biotechnologies.

My general thesis is that the headline disjunction is not an exclusive disjunction. As far as I can see, genetic engineering might be used in qualitatively different directions, ecologically as well as evolutionary. The two extremes are genetic pollution and second order evolution. What I will try to do is to find a criterion for distinguishing a second order evolution perspective from a genetic pollution perspective in relation to genetically modified animals. In order to keep the discussion within reasonable bounds, I will assume that the following two preconditions are fulfilled:

1) The genetically modified animals do not cause any reduction in the existing biodiversity, and
2) The animal welfare is not reduced, compared to that of not modified relatives, according to the most comprehensive animal welfare parametres.

The genetic pollution perspective and the second order evolution perspective will be examined separately. In the discussion of second order evolution, my point of departure will be Darwin's thesis, that the development of social instincts and moral feelings is a decisive factor within evolution. My main argument is, that it may be more fruitful for the preservation of the authenticity and integrity of nature to stimulate the development of social instincts of all relevant creatures, instead of simply preserving the genomes unmodified. My presumption is, that in so far as this stimulation is fruitful (without conflicting with the two preconditions mentioned above), one should regard the modified animal as an integrated part of the natural biodiversity and, accordingly, as equally worthy of our concern, protection and preservation.

The genetic pollution perspective

If 'genetic pollution of nature' shall be ascribed any meaning at all, some criteria have to be set up. I think that the following criteria (or something similar) could be accepted by everybody, who sees the term 'genetic pollution' as meaningful. An animal (or more generally: an organism) causes genetic pollution, if it:
  1) has been subject to a species boundary transcending genetic engineering
  2) escapes into uncultivated areas in spite of human attempts to keep it inside a laboratory, a stable or a fence
  3) mates with a not domesticated /wild animal (or more generally: an unmodified organism)
  4) succeeds in transferring the genetic modification to next generation

I assume that these conditions are *sufficient* conditions for a characterization of genetic pollution. Whether all four of them are *necessary* conditions, too, this is a question which I will leave unanswered. An animal which fulfills the conditions represents as well as causes genetic pollution. Thus, in the following discussion I shall presume that it can be seen as a paradigm of a genetic pollutor.

The extension of this interpretation of 'genetic pollution' is quite narrow, I admit. It might very well be possible to bring forward some good arguments for extending the interpretation by analyzing and relating it to broader criteria. One could, for instance, argue that genetic pollution is already present, as soon as there are genetic changes induced by humans on organisms remaining within laboratories, stables and fences, or genetic changes caused by selective breeding. My main reason for not using these broader criteria is the presumption, that this might put us on a wrong track, where we would be tempted to overlook the factors, which I see as decisive for preventing the reductions of authenticity and integrity of nature. So let us stay with the narrower definition, and turn to the second order evolution perspective, from which these factors become visible.

The second order evolution perspective

The second order evolution perspective can shortly be characterized as 'Develop and let others develop! Evolve and let evolve!' Or even stronger: 'Evolve *through* letting others evolve!' A basic point in this slogan is the demand, that *if* we take upon ourselves a responsibility for promoting and accelerating the course of evolution (and this is what we do by accepting genetic engineering), *then* the most appropriate criteria to use when separating good and bad developments, is to let the creatures develop the characters and capacities most valuable in an evolutionary perspective, at least to the same degree as they would have done without human interference.

Immanent to the second order evolution perspective is the presumption, that humans potentially represent a provisional pinnacle of creation, being measure as well as motor of the actual evolutionary development. The most spectacular sign of this role is the construction of complex, powerful technologies. The fundamental question is, however, whether it is by virtue of the capacity itself of constructing sophisticated technologies, that humans ought to be most satisfied with themselves as motors of evolution, or whether it is, rather, by virtue of reflecting upon the role and longsighted and wideranging consequences of the potential technological applications. In the latter case, humans take upon themselves the role of evaluating the evolutionary processes according to a complex, ethically ambitious scale, and of critically selecting among the many possible technological applications only those which will enhance an ethically sound development. In short, the question is whether it is in virtue of the development of our ethical capacities, not least of our responsibility, care and compassion, rather than by the development of our technological skills, that we are most worthy as motors of evolutionary change.

According to my best judgement, the realization of the most promising perspectives of these two dimensions of development depends on our success in not seeing them as exclusive alternatives. A more fruitful route to take is to try to combine the two by, on the one hand, letting the technological phantasy and creativity challenge our ethical sense, when attempting to integrate the new technological potentials into our (thus gradually changing) ethical horizon, while, on the other hand, letting those potentials be realized only, which we find acceptable from our present point of view. Without this double perspective, technological applications could only be seen *either* as evolutionary obstacles *or* as unchallengable motors of evolution.

In order to develop the second order evolution perspective further, let us consult the father of evolutionary theory, Charles Darwin. In his *Descent of Man*,[1] Darwin claims that humans as well as the socalled lower animals possess social instincts, which give rise to sympathies and altruistic wishes to help fellow creatures. As humans have developed intellectually, they have become more and more able to recognize future consequences of their acts, and as their sympathies have expanded and come to embrace all kinds of humans, including those with physical or mental handicaps, and later on socalled lower animals, too, their moral level has risen correspondingly.

Darwin emphasizes that the moral capacities are rooted in the social instincts, which are common to humans and several other animals, whereby opening the door for social development of nonhuman creatures. If social or moral development is regarded as a fundamental motor, maybe even the very *telos* of evolution, I find it quite natural to say also, that the social and moral development of all organisms, in accordance with their means and abilities, is extremely important. Not just in itself, but also as a precondition for the, from this point of view, most sophisticated evolutionary process of all: human moral development.

I find it most naïve to imagine that humans can develop morally, independently of their treatment of nonhuman creatures. I do not just think in terms of physically cruel treatment and psychological cruelty, however, but include treatment, that prevents the creature from expressing its social instincts and sympathies for relatives, particularly its youngs, and for other creatures. Prominent philosophers like Thomas and Kant have argued that there is a brutalizing impact on the agent who treats animals with cruelty, whereas only few have noticed the second kind of impact. From my point of view, however, there is an important qualitative difference between the two cases. In the first case the considerations for the animals can be expressed in the terms of traditional animal welfare ethics, and can influence our "first order evolutionary" traits, primarily our immediate moral capacities like care and compassion. In the second case, however, what is at stake is our capacity to represent a secondary evolution, i.e., to assist the evolution.

I shall conclude that a necessary condition for categorizing an application of genetic engineering as potentially expressing a second order evolution, is, firstly and evidently, that the first order evolution demands be fulfilled, and secondly, that at least it should not *prevent* the organisms from developing and expressing their spontaneous social instincts and moral feelings, particularly not their care for their youngs and other fellow creatures. Such preventions represent obstacles to evolution, following the Darwinian interpretation referred to above. Whether they represent genetic pollution, too, depends on their succesful mating with not domesticated animals.

Biotechnological impacts on the present and future biodiversity

It will inevitably have decisive impact on our use of the biotechnologies, whether the genetic pollution perspective or the second order evolution perspective will become dominant. Retrospectively, it turns out that until recently the latter has, implicitly, been dominant during this century. Quick and powerful biotechnologies have not been available until the latest few decades. Nevertheless, innumerable organisms resulting from selective breeding and cultivation have become new partners in the biosphere.

This process has been so relatively slow and stepwise that people have gradually become accustomed to the changes of both genotypes and phenotypes, even though several of the changes are quite visible. Among the most conspicuous changes of farm animals and pets, one could mention the fourfold increase of milk production per dairy cow, causing a high degree of mastitis, or the turkeys with so enlarged breast muscles that they cannot mate without artificial insemination, or dogs which have developed chronic eye inflammation ("appealingly sad eyes"), breathing difficulties and whelping problems, necessitating routine surgical operations.

These and a series of other gradually changed organisms have been integrated into contemporary biodiversity, roughly speaking, without critical questioning. Seen in the light of this, it is quite remarkable that the *Convention on Biological Diversity*[2] gives no direct answer to the question concerning the possibility that genetically modified organisms could increase biodiversity. Not even a single paragraph indicates that they could. The aim of the convention is, evidently, to *preserve*, i.e., to hinder the reduction of present biological diversity, including the cultivated and domesticated organisms that can be used sustainably; it does *not* mention the possiblity of increasing diversity artificially. Organisms and "alien species" (Article 8.h), resulting from application of biotechnology, only represent risks, as they can "threaten ecosystems, habitats or species" (ibid.), and should accordingly be controlled or prevented. The ideal is to "provide the conditions needed for the compatibility between present uses [of modified organisms (MS)] and the conservation of biological diversity and the sustainable use of its components" (Article 8.i).

Nevertheless, I shall attempt to expand the field of interest beyond that of the convention. From my point of view, it is precisely in relation to the applications of biotechnologies, which are compatible with the conservation of the present biodiversity and the sustainable use of its components, that the question concerning genetic pollution versus second order evolution perspective represents the greatest challenge. As long as genetically modified organisms are seen first and foremost as a threat to biodiversity, this threat is what counts, and our two perspectives are of secondary importance only. However, once the protection of present biodiversity is secured to a satisfying degree (and this is in itself an extremely difficult task), or more likely: even before then, another question arises, namely under which conditions new and artificially modified organisms can be introduced into the biosphere.

My fairly ambitious aim is to propose a single criterion for a reasonable application of the term "increase of biodiversity" or even, if possible, "enrichment of biodiversity". In the latter case the quantitative dimension of stability needs a qualitative supplement, for which I cannot find a better expression than the so far unsurpassed dictum of the grand old environmental philosopher Aldo Leopold: "A thing is right when it tends to preserve the integrity, stability and beauty of the biotic community. It is wrong when it tends otherwise."[3]

As I see it, the Convention on Biological Diversity concentrates on the preservation-of-stability-of-the-biodiversity dimension, whereas I want to focus on the two supplementary, overlapping dimensions, namely those of, preservation-of-integrity, to which I add authenticity, and preservation-of-beauty. I find it pressing to include these supplementary dimensions for two reasons. Firstly, I presume that some of the earlier mentioned examples of genetically modified farm animals and pets, developed via traditional selective breeding during this century, and apparently integrated into present biodiversity, would not have been accepted, according to the Convention, if these dimensions had been parts of a criterion for acceptance. And secondly, I foresee that several of the planned, dreamed-of and not-yet-imagined genetically modified animals will pass the criteria of the Convention, at least as "alien species", safely controlled within laboratories, stables or fences,- even though they are probably more radically changed than the previously mentioned examples of changes caused by selective breeding.

On the other hand, I cannot accept a definition which is so restrictive that it excludes a priori the possibility of using genetic engineering, which transcends species boundaries, but still respect the second order evolution criteria. I regard the genetic pollution point of view as secondary to that of a second order evolution in a preservation-of-authenticity-and-integrity-of-nature perspective. From the Darwinian interpretation, which I find to give us an optimal basis for ascribing content to integrity and authenticity, I draw the conclusion, that it might be more essential not to reduce the animal's possibilities for developing social instincts and sympathies, than to keep its genome unchanged.

Consequences in practice

So far genetically modified farm animals have had very little success. Great expectations were attached especially to presumed possibilities of increasing the meat and milk production by inserting genes for growth hormone. The Beltsville swine, which developed several severe welfare problems without increasing weight, represents a well known failure. On the other hand, Canadian genetic engineers have succeeded in transferring extra growth hormone genes into Pacific salmons, whereby making them grow to 10 to 37 times the weight of their not modified relatives, apparently without welfare problems.[4] But hardly without problematic ecological consequences, if they were permitted access to the ocean (for which reason they would probably not live up to the criteria of the Convention on Biological Diversity). In spite of the present failures, however, there is little doubt that some researchers expect important potentials in applying genetic engineering on farm animals, too. The basic ambition is to take a series of specific, highly appreciated qualities from several species, and combine them into one single species. If this becomes reality, we will have to find criteria to use in the process of selection.

My earlier stated conclusion concerning development of social instincts has become particularly relevant in relation to some recent promising implementations of genetic engineering on animals. What I have in mind is the construction of transgenic bioreactors, i.e., animals which produce human medicine (or other useful substances) together with the production of milk, like the alfa-1-antitrypsin producing sheep Tracy and the factor IX producing Scottish sheep. As long as the animals are kept safely within laboratories, stables or sufficiently strong fences, they do not represent a problem of genetic pollution, nor can they be seen as a risk or a threat to present biodiversity. Furthermore, as long as no unintended side effects occur, they do not represent any obstacles from an animal welfare ethics point of view either, as the genetic engineering has taken place at the level of embryo, before the development of the nervous system. From a second order evolution point of view, however, the evaluation is more complex. For instance, one of the most common consequences of constructing such bioreactors is that the animals are prevented from expressing their social instincts, particularly from nursing and taking care of their youngs. In the case of the medicine producing sheep, the bitter ironi is, that it is a necessary precondition for their role as bioreactors, that they bear and give birth to youngs, whereas this very role make them unable to nurse their youngs. Whether this prevention from expressions of social instincts is due to the change of chemical composition of the milk, caused by the genetic modification, or only to financial considerations, such interferences cannot be accepted as a second order evolution representation, but must rather be seen as obstacles to the course of evolution.

However, these examples do not exclude that some applications of species boundary transcending genetic engineering on animals could be accepted within my interpretation of Darwinian theory. The authenticity and integrity of animals is not a priori violated by operations of genetic engineering. Engineering is acceptable, if it does not reduce, but rather stimulate a further development of the social instincts of the modified animals.

Conclusion

If it is our ambition to let our acts and operations adapt to a reasonable interpretation of second order evolution, then we have to change our biotechnological innovations of animals accordingly. This means, that we must strive always to let the animals develop their social instincts, care and mutual aid toward their fellow creatures (as emphasized by the Russian anarchist P.A. Kropotkin), at least to the same degree, and preferably even higher than they would have developed without our interference. Whether the animals live in a laboratory, a stable, on a field or in wilderness is of no relevance in regard to this.

The fulfilment of this demand may be compatible with some interferences with the genomes of the animals. Interferences, i.e., which can be categorized as polluting factors according to the previously stated definition of genetic pollution. What is decisive, however, is that we remain within a second order evolution perspective, i.e., that we give the highest priority to the social instinct considerations before accepting any concrete case of biotechnological interference. Only this way can the main dimensions of the integrity and authenticity of the animals be respected. Accordingly, I see no reason for not letting such animals increase, possibly even enrich the biodiversity.

Notes

1 Darwin, C.: *Descent of Man*. I have used the first Danish edition: *Menneskets Afstamning*, København: Gyldendal 1873, pp. 386f.
2 *Convention on Biological Diversity*, United Nations Environment Programme, Na. 92-7807, 1992.
3 Leopold, A.: *A Sand County Almanac*, N.Y.: OUP 1966, p. 240.
4 Delvin, R., Yesaki, T.Y., Blagl, C.A., Donaldson, E.M.: "Extraordinary salmon growth", *Nature* 371, 1994, pp. 209-210.

Biodiversity and the importance of the legal framework

Helle Tegner Anker

Introduction

Implementation of the biodiversity objectives – as laid down in, inter alia, the Biodiversity Convention – amongst many factors depend on the legal framework at international, regional and national levels.

The purpose of this paper is not to provide a comparative analysis between different legal systems, but rather to call attention to the importance of understanding some basic elements of legal systems and the legal barriers or opportunities for implementing biodiversity objectives. This includes questions of how legal systems may encompass considerations of the so-called intrinsic value of the biological resources and how traditional concepts of law, such as the concept of property rights and the rules of law and order, may be compatible with biodiversity protection issues. It is pointed out that potential areas of conflict between legal traditions and environmental goals often require an open and informed decision-making – and that the legal framework plays an important role in this respect.

The concept of biodiversity

Biodiversity is a very complex concept and the concept has received an increasing amount of attention in recent years. The signature of the Biodiversity Convention at the Rio Summit in 1992 came as an international response to such biodiversity considerations primarily aimed at reversing the increased extinction of species. The Biodiversity Convention lays down a three-fold objective of:

- biodiversity conservation,
- sustainable use of the biological resources and
- fair and equitable sharing of benefits arising out of the utilization of the genetic resources.

While the last of these objectives perhaps has been the most controversial in the negotiations of the convention,[1] the two former objectives may be of a more fundamental and far-reaching character.

Various efforts have been made to clarify the concept of biodiversity, particularly by natural scientists but also by economists.[2] Some of the key elements deriving from these efforts have been species richness and ecosystem resilience indicating the possibilities of arriving at inventories and quantifiable values of biodiversity. Yet, the difficulties of establishing inventories and quantifying values of biological diversity seem to be overwhelming, not least considering the different types of values, *inter alia*, the question of intrinsic versus instrumental value.[3] Also the somewhat "contradictory" biodiversity definition (diversity within species, between species and of ecosystems) is likely to make it difficult to specify the exact meaning of biodiversity in a traditional scientific manner.

From a legal perspective – and perhaps also from a broader humanistic and social science perspective – it appears that biodiversity in general touch upon the relationship between man and nature including both inter- and intergenerational aspects. It has been argued, that the extensive species loss suggests a *"need to engage in a fundamental reassessment of the relationship between human society and the natural world."*[4] A key concern of biodiversity protection seems to be to avoid, minimize or reverse adverse effects of human activity. Following this perception sustainable use of the biological resources becomes very central to biodiversity protection. Thus, efforts to secure sustainable use must supplement more traditional measures of nature conservation. This implies that the legal and regulatory means of implementing biodiversity objectives penetrate into a broad range of different legal and regulatory systems – and not only concern traditional nature conservation law.[5]

The legal framework

The legal framework may also be considered a rather complex concept operating at several levels. The legal framework does not only refer to specific pieces of legislation at international, EU and national level. Policy documents and the like, *inter alia*, so-called action plans, may also be part of the legal framework stating objectives, principles and possible means of action. In an international perspective such documents are often termed 'soft law'.[6] Furthermore, unwritten principles of law, e.g. principles of administrative law, are included in the concept. Clarification or interpretation of such elements by case law is of course also included.

The legal framework does not only consist of substantive and procedural requirements. Issues of organizational structures, competencies and the choice of legal and regulatory instruments are part of the legal framework. The legal framework thus constitutes part of what one could call decision-making structures, creating legal norms for human behaviour. The legal norms or the legal framework interact with and constitute important parts of the complex structures and cultural context of any society.

In the course of implementing environmental objectives, *inter alia*, the objectives of biodiversity protection, it must however be realized, that the legal framework or certain elements of it may to a varying degree facilitate implementation. In some case, parts of the legal framework may even pose barriers or deficits to the implementation of environmental objectives. In fact one could in general speak of a balance between environmental objectives and principles on the one side and traditional principles of law (e.g. principles of law and order, property rights and competition) on the other side. Such barriers and the importance of a proper balance may be overlooked both in relation to the formulation of new rules and in the application of law, as they are often "hidden" in the decision-making structures. But, as mentioned above, the character and the importance of such legal barriers very much depend on the particular cultural context of which they are part.

Basic elements and principles of law

The traditional principles of law are of course not identical in different legal systems. Yet, there appears to be a common core at least in most Western countries stemming from the traditional purposes of law to protect the individual citizen against the misuse of public powers and to solve conflicts between individual rights and interests of the citizens. The meaning of such principles, e.g. law and order and private property, is, however, far from precise and not static. The content of the principles develops over time and adapts to societal circumstances. Yet, the principles are closely related to the concept of justice and as such important for the understanding and acceptance of law and at the end of the day for implementation of law in practice. This is particularly true in relation to environmental problems caused by the cumulative effects of many small actions, *inter alia* effects on biological resources caused by the practices in farming, forestry, tourism, transport or energy consumption in relation to which traditional means of law enforcement and 'command and control' may be difficult to apply.

The question of rights and interests

One interesting feature of the legal framework and the ability to safeguard and protect the biological resources is the question of rights and interests. In a legal terminology rights and interests are very closely related concepts. But, normally only human beings can be holders of *rights*, while other beings or assets can merely be ascribed rights or values indirectly, relative to man. The term *interests*, thus, may be said to represent values that are to be safeguarded by man.

The idea of ascribing rights to natural objects was advocated by Christopher Stone in 1972.[7] Stone suggested the creation of a guardianship approach in order to make the rights status applicable to the legal system and to the courts. It may, however, be argued that a rights-based approach implies a circumscription to human values and perceptions of nature, and does not necessarily express the values of nature as such.[8] One could also question the ability of the legal system to handle conflicts between natural objects, if a rights-based approach is applied. Yet, it is likely to be very difficult for many legal systems to accept a rights-based approach, at least when it comes to natural assets or objects that perhaps are less conspicuous than, e.g., whales and other "popular" species. Thus, in relation to biodiversity issues – that is the protection of the biological resources as such – most legal systems are probably more likely to accept, and perhaps also benefit from, an interest-based approach rather than a rights-based approach.[9]

Generally, it could be claimed that a rights-based approach may "hide" the human values and perceptions, while an interest-based approach rather may clarify the different interests involved under the presumption of an open and informed interest-balancing process. Critics have also been raised on the point that ascribing values (or rights) to peoples, animals or the environment "*does not tell us what that value should be or how it should be weighed against other values or 'rights'.*"[10] Yet, the legal rights status advocated by Stone was not an absolute right (as only very few rights are absolute) and Stone declared: "*What the environment must look for is that its interests be taken into account in subtler, more procedural ways.*"[11] Thus, in essence there may be no disagreement on the need for considering the values of nature and environment also in the legal systems, and the main problem can be identified as the one of implementation.[12] In this respect, implementation of an interests-based approach may be more feasible than implementation of a rights-based approach, although as pointed out by Stone "*rights introduce a flexibility and open-endedness that no rule can capture.*"[13]

Most Western legal systems are directed towards the protection of the individual rights and interests of the citizen. In environmental law many interests of the public or the society, *inter alia*, clean air, water and soil, are safeguarded by means of public environmental law, e.g., via authorizations and emission standards. However, such interests are normally also characterized by their economic nature or value. The situation appears to be somewhat different in relation to biodiversity interests that are not directly related to individual persons or to the immediate sphere (environment) of such persons. Is it, at all, possible to determine the interests of nature as such, and if so, who is able to determine these interests? The uncertainty attached to non-economic interests or values may also, in traditional economic terms, lead to a discounting in promoting the more present (economic) interests.

The legal protection of interests can be divided into a direct protection by substantive requirements, and a more indirect or procedural protection secured by requirements of letting the different interests 'participate' in decision-making (e.g. by assessment procedures, access to justice/administrative proceedings, access to information and other procedural requirements).[14] In order to protect the largest number and variety of interests, the legal framework must appeal to an open and informed decision-making. Furthermore, legal protection of vague and imprecise interests, such as the interests of future generations and nature, must be characterized by duties of reflection, consideration and responsibility. In other words, a proper balancing of all interests involved needs to be promoted by a legal framework expressing the *'virtues of deliberation'*.[15]

Another more direct means of recognizing the obligation to consider the different values of nature may be the use of constitutional provisions.[16] A similar expression of such an overarching principle follows from the preamble of the Biodiversity Convention, which underlines the need to be *"conscious of the intrinsic value of biological diversity."*

Private property

Traditionally, private property is one of the well-protected individual, economic rights of most legal systems – often referred to in constitutional provisions. Private property rights of real estate are also from time to time presented as one of the most serious obstacles to implementation of biodiversity objectives. The reasons for this may vary from one legal system to another, but they particularly relate to the questions of enforcement and costs. Biodiversity protection is especially confronted with private property rights by restrictions on land use and development in connection with species and habitat protection. Not only traditional nature conservation initiatives, but also requirements for sustainable (land) use may impose such restrictions, *inter alia* relating to farming and forestry practices.

Two central questions are, to which extent may changes in land use be (en-)forced upon the landowner and which costs is the landowner expected to bear? In a Scandinavian perspective,[17] it appears to be clear that the landowner may be obliged to refrain from certain uses, but that he cannot be forced to take more active steps (*inter alia,* nature management) unless an agreement has been reached.[18] But the dividing line between active and passive steps is not very clear as 'passive' restrictions may force the landowner to take active steps. In a Scandinavian context it also appears that the costs related to restrictions in future land use must be borne by the landowner, while restrictions in existing land use are more likely to lead to compensation. But also in this respect the borderlines are rather unclear. In the US the central distinction is between the exercise of police power and the exercise of eminent domain (taking land for public uses). The former exercised with the purpose of controlling public nuisances and without compensation.[19] Although, the legal and cultural context may provide different criteria, it seems clear that the restrictions directed towards biodiversity protection are difficult to categorize.[20]

However, private property rights are not absolute nor static rights. The content of private property rights varies over time and within the cultural context. It is widely held that private property rights serve a social function and that ownership consists of both rights and responsibilities.[21] The responsibility aspect may be expressed in different ways, one of which is the acceptance of different kinds of interferences with private property.

Most legal systems accept such interferences with private property. The principles of neighbour law have been extended through a massive body of environmental law to safeguard the interests of the public. Balancing of individual interests against public interests in a clean environment is central to environmental protection law. This is also the case in relation to nature conservation law, although the private property rights of the landowner traditionally have been compensated in the case of conservation orders, establishment of wildlife reserves, national parks etc. – again for the (immediate) benefit of the public. In relation to biodiversity, however, the interests of the biological resources are much more imprecise, vague and intangible. Thus, biodiversity interests are likely to be undervalued in the interest-balancing process. The US Endangered Species Act (ESA) provides one example that this is not necessarily true in a legal sense by providing for a "strict" protection of any species listed according to the Act. However, after the famous Snail Darter Case, the ESA was amended by the introduction of an exemption process and a cost-benefit analysis.[22] Yet, also the species by species approach in the ESA has been termed as inadequate in the pursuance of broader biodiversity objectives.[23]

Due to the fundamental, but not absolute nor precise position of property rights in our legal systems, they cannot be overlooked or ignored in environmental matters. Neither does a conflicting approach setting environment and private property fundamentally at odds[24] appear to be feasible in the long run. In Denmark, a steady but slow increase in what we call compensation-free regulation within nature protection law has been testing the limits of private property rights during recent years. But this has been done rather 'discretely' without proper deliberation on the real subject of private property rights. Although it is very difficult to point out specific pieces of legislation which violate private property rights, the mere amount of legislation and regulation with potentially interfering consequences has recently caused much discussion in Denmark, particularly amongst farmers. This is also due to the use of more indirect regulatory instruments, such as planning and area designations, creating a great deal of uncertainty among farmers. In some cases, counterproductive activities – such as ploughing up meadows in order to avoid falling within the scope of nature protection rules – have been reported. Such activities must be characterized as very unfortunate consequences of not facing the core issues at stake.

Generally, it can be said that biodiversity protection and, in particular, the sustainable use of the biological resources requires a minimum of understanding and acceptance of the legal framework. Thus, the balance between rights and responsibilities of landowners is very important – not least in relation to biodiversity issues. In this respect both the formulation and the application of law play a key role. Also the choice of legal and regulatory instruments may emphasize the responsibility aspect. Such instruments are likely to be found within what has been called market-based or 'bottom-up' instruments creating incentives for the environmentally sound management of resources.[25] Such instruments may include agreements, nature restoration, land allocation and subsidy-schemes, but they may also imply the removal of disincentives.[26] The Biodiversity Convention also calls for the adoption of incentive measures for the conservation and sustainable use of components of biological diversity, cf. article 11. Although many of these instruments may be characterized as economic instruments, they all need to function within a legal framework.

While the Biodiversity Convention does not explicitly touch upon the issue of the private property rights of landowners, other property issues are reinforced in the Convention. This is particularly the case in relation to access to genetic resources, where both the sovereign rights of States and the so-called intellectual property rights of knowledge are clearly recognized, cf. article 15 and 16. By putting these types of property rights forward as means of biodiversity protection, the element of responsibility is strongly emphasized. Yet, one could argue that this kind of responsibility is closely related to economic interests and to the principle of equitable sharing of benefits. In relation to private landowners, the responsibility inherent in property rights should rather reflect a general considerability towards nature and future generations.

Law and Order

A third – but not separate – basic element of most Western legal systems is the protection of the individual citizen against the misuse (or abuse) of State power – expressed in the principles of *law and order*, including what we in Denmark call the principle of legal certainty (rule of law) and the principle of proportionality. The principle of legal certainty is, like the principle of private property rights, a vague and imprecise principle referring to the procedural elements of overseeing and influencing the legal framework and the application of law, and to the substantive elements of securing the individual citizen a just and fair position in the legal system. The principle of proportionality relates to the principle of choosing the least interfering means of implementation, thus implying some sort of balancing against economic interests.

Yet, both principles may potentially conflict with the *precautionary approach* which characterizes biodiversity issues. The precautionary approach recognizes the often significant degree of uncertainty in relation to the impacts on biological resources of various human activities. One could argue, that the significant degree of uncertainty makes it very difficult to determine the necessity or the proportionality of certain interferences, and also, to oversee the legal position of the individual. Thus, again a sensitive balance exists and must be taken into consideration in the same ways as mentioned above in relation to the issues of private property rights.

Although, the precautionary *principle* is not explicitly laid down in the articles of the Biodiversity Convention, it is reflected in the preamble of the Convention, which states: *"..noting also that where there is a threat of significant reduction or loss of biological diversity, lack of full scientific certainty should not be used as a reason for postponing measures to avoid or minimize such a threat.."*

While the precautionary principle appears to be fairly narrow in scope and directed towards specific threats against biodiversity, a broader precautionary approach is also reflected in the preamble of the Convention: *"..as (conscious of the different values) and conscious also of the importance of biological diversity for evolution and for maintaining life sustaining systems in the biosphere."*

Also the notion of the interests of future generations is in line with a precautionary approach. As we cannot at present identify the needs and aspirations of future generations, a precautionary approach appears to be the only way in which the present generation can safeguard the interests of future generations.

The precautionary approach express a responsibility of facing and deliberating on the elements of uncertainty in any aspect of life. Essentially, a precautionary approach requires a careful and open balancing of different interests involved, including the aspects of uncertainty. The openness of such a balancing process is important, as openness and transparency are prerequisites for a responsible decision-making process. In relation to the legal framework, risk assessments and other forms of assessment are designed not only to deal with uncertainty, but also to establish the basis for an open and transparent decision-making process. Other procedural instruments, such as a reversal of the burden of proof, may also facilitate a precautionary approach as part of the legal framework.

Concluding remarks

The intention of this presentation has been to illustrate the importance of realizing and considering the role of the legal framework with particular focus on the implementation of biodiversity objectives. I have tried to draw out three potential areas of conflict between basic elements of law and biodiversity objectives. The point is, that such potential conflicts should be dealt with in a proper manner considering the complex interest balancing in decision-making processes. If they are not properly deliberated upon, such elements may create legal barriers or deficits to the implementation of biodiversity objectives. From a Danish perspective, I can say that these issues are seldom deliberated upon properly either in the drawing up of new rules or in the application of law.

Basically, the conflicts illustrate the ever-existing balance between environmental goals on the one side and other (often economic) societal goals on the other side. Especially in relation to the conflict between economic activity and species protection it has been argued that (only) *clarifications of the structure for resolving conflicts may enable us to reach better and more widely accepted decisions in the future.*[27]

One of the main reasons why this aspect is particularly important in relation to biodiversity protection, is the extremely broad-ranging character of biodiversity issues. It is not possible to confine biodiversity protection to traditional nature conservation. The biodiversity thought needs to be integrated into nearly all kinds of sectoral activities and into the everyday life of the individual citizen. In this respect one cannot rely on the traditional measures of 'command and control' and law enforcement. Yet, one must rely on the general understanding and acceptance of law and hereby the responsibility of the individual actor. The concept of justice thus becomes central to the implementation of biodiversity objectives, and the concept of justice is strongly related to the above mentioned basic elements of law.

Open and informed discussion of the potential areas of conflict appears to be a prerequisite in this approach and the legal framework must be based on such discussions. Furthermore, the legal framework should facilitate open and informed decision-making at all levels, e.g. by procedural requirements. In this respect environmental (impact) assessments in a broad sense may play a key role based on a consensus approach. Hereby, the legal framework may encompass a more soft, integrative approach emphasizing the inevitable balance between nature protection and other societal goals.

Notes

1   Hendrickx, F., V. Koester and C. Prip: 'Convention on Biological Diversity. Access to Genetic Resources: A Legal Analysis', *Environmental Policy and Law*, vol 23 no. 6, 1993, pp. 250-258.
2   For example C.A. Perrings, et al: *Biodiversity Conservation*, Dordrecht: Kluwer Academic Publishers, 1995. See also Veit Koester: 'The Biodiversity Convention Negotiation Process', in Basse, E.M. (ed.) *Environmental Law. From International to National Law*, Copenhagen: GadJura, 1997 pp. 205-258.
3   See for example Michael Bowman: 'The Nature, Development and Philosphical Foundations of the Biodiversity Concept in International Law', in: Bowman and Redgwell (eds.): *International Law and the Conservation of Biological Diversity*, London, Kluwer Law International, 1996, p. 15.
4   Kellert, Stephen R.: 'Social and Perceptual Factors in the Preservation of Animal Species', in: Bryan Norton (ed.): *The Preservation of Species*, Princeton, New Jersey: Princeton University Press, 1986.
5   See also Philippe Sands: *Principles of International Environmental Law Vol. I*, Manchester: Manchester University Press, 1995, p. 450: '...*requiring a comprehensive approach to regulation of a broad range of human activities*'.
6   See Patricia W. Birnie and Alan E. Boyle: *International Law & the Environment*, Oxford: Clarendon Press, 1992 (1994), pp. 26-30.
7   Stone, Christopher D.: "Should Trees Have Standing – Toward Legal Rights for Natural Objects", *Southern California Law Review*, vol. 45, 1972, pp. 450-501.
8   From a philosophical point of view J.B. Callicott argues, that 'species rights' symbolises the widely shared intuition that nonhuman species possess intrinsic value. Yet, Callicott regrets the use of the concept of rights as a way of expressing moral considerability. J. Baird Callicott: 'On the Intrinsic Value of Nonhuman Species', in Bryan Norton, *op. cit.*
9   In relation to the rights of future generations, they may more easily fit into the legal framework, which has also been the case in some countries, *inter alia*, the Philippines. See IUCN: *Watching the Trees Grow: New Perspectives on Standing to Sue for Environmental Rights*, IUCN CEL Philippine Group, 1995.

10 Birnie & Boyle *op. cit.*, p. 189.
11 Stone *op.cit.* p. 483.
12 Birnie & Boyle *op. cit.*, p. 213.
13 Stone *op. cit.*, p. 488.
14 Also economic and other indirect instruments may be used as part of the legal framework to influence decision-making.
15 See Mark Sagoff: *The economy of the earth. Property, law and the environment*, Cambridge: Cambridge University Press, 1988, p. 17. See also William Snape (ed): *Biodiversity and the Law*, Washington D.C.: Island Press, 1996, p. 219.
16 See Rodger Schlickeisen: 'Epilogue', in William Snape *op.cit.*
17 See Gabriel Michanek: "Principer för ersättning vid rådighetsinskränkningar i vissa stater", *Miljörättslig Tidskrift* 1996:1, pp. 1-45, IMIR for a comparison of private property rights in relation to biodiversity protection in some of the Nordic countries and the US.
18 Westerlund, Staffan: 'Rättsliga Floravårdsfrågor', *Miljörättslig Tidskrift* 1994:1, pp. 78-97, IMIR.
19 See Robert L. Carlton: 'Property Rights and Incentives', in Bryan Norton (ed) supra note 4.
20 Robert L. Carlton *op. cit.*
21 See Juergensmeyer, J.C.: 'The American Legal System and Environmental Pollution', *23 UN. OF FLO. L. ReV*. 439 (1971), see also Gabriel Michanek: 'National Protection of Biological Diversity', in Basse, 1997 *op.cit.*, p. 310, with reference to the German Constitutional rule: *'Eigentum verpflichtet'.*
22 Flournoy, Alison: 'Beyond the Spotted Owl Problem: Learning from the Old Growth Controversy', *Harvard Environmental Law Review*, vol. 17, 1993, pp. 261-332.
23 Flournoy *op. cit.*.
24 Olson, Todd G.: 'Biodiversity and Private Property: Conflict or Opportunity', in Snape 1996 *op.cit.* p. 69.
25 See also Minna Gillberg & Håkan Hydén: 'Law as a Safety Belt and as Enforcer', in Swedish Council for Planning and Coordination of Research: *Challenges in Environmental Human Dimensions*, Stockholm, Report 1996:10.
26 See Brian J. Preston: 'The Role of Law in the Protection of Biological Diversity in the Asia-Pacific Region', *Environmental and Planning Law Journal*, vol. 12, 1995, pp. 264-277.
27 Flournoy *op. cit.*, p. 323.

Indigenous Peoples
and the Protection of Nature

Common knowledge and resource conservation, globally and locally

Paul Richards

Common-sense beliefs

According to Scott Atran, the philosopher G. E. Moore considered common-sense beliefs to be universal propositions commonly held and, so far as is known, always believed. He cites as examples belief in existence of material objects, or that people (normally) have thumbs. Moore distinguished common sense from popular beliefs that are widespread but erroneous, or popular but unreasonable opinions.[1]

Common sense and conservation

Let me now introduce the main point to be developed in the present paper: effective conservation of biological resources implies, politically, either enforcement by superior authority or a mobilization of popular will. This paper expresses a preference for mobilization of popular will.

I start from the assumption that popular will for conservation would be better based on scientific truth, than upon Moore's "*widespread but erroneous, or popular but unreasonable opinions.*"[2] Ergo, conservation requires public understanding of science – either "top-down" by careful 'popularization', or "bottom-up", by establishing connection with the world of common sense.

I propose that building upwards from a platform of common sense is a good strategy, for two reasons. Cognitively, common sense can be shown to be "an independent but faithful ally of science."[3] Second, to treat common sense with respect is consistent with, and therefore reinforces, basic commitment to democratic mobilization of the popular will.

Conservationists, therefore, need to understand common sense. Key questions (following Atran) are:
– What constitutes common sense concerning plants and animals, water, earth and air, fertility and reproduction?

- How is this knowledge formed and sustained?
- Where does common sense end and specialist knowledge or popular nonsense begin?
- In what way, if at all, can universal human capacity for common sense be drawn upon as a foundation or platform for popular participation in bio-resource management?

Science and common sense: inductivism and anti-inductivism

Various views are held about the relationship between common sense and science.

Wolpert[4] has recently forcefully re-stated the notion that 'science' and 'common sense' are polar opposites. For Wolpert, science is counter-intuitive The title of his 1992 book makes his point with exact economy – *The Unnatural Nature of Science*.

Wolpert's book endorses what Scott Atran terms the anti-inductivist position in philosophy of science. Atran characterises the standpoint of the anti-inductivists thus: "*common sense is hopelessly mixed with symbolism ... science truly grows only to the extent that it purges thought of this confusion ... scientific knowledge is not achieved by sifting through and refining customary thought, but by discarding the whole mess outright.*"[5]

Inductivists, by contrast, are happy with the idea that many of the best ideas in science depend in some way upon widely-shared human understandings based on experience. Where anti-inductivists see an unbridgeable gulf between the abstractions of science and everyday experience inductivists sense an unbroken if tangled web of ideas, knowledge and activity. Even the counter-intuitive, Atran notes, takes the intuitive as its point of reference ("*What makes counter-intuitive ideas understandable at all is that they remain rooted in common-sense intuitions, however remotely*").[6]

Anthropology and ethnoscience

Anthropologists have contributed to the inductivist/anti-inductivist debate by taking positions on so-called 'primitive thought'. 'Primitive thought' is a label for the knowledge and beliefs of early or non-literate human groups.

The British 19th century anthropologists Frazer and Tylor considered primitive thought to contain a kernel of rationality, however thick the shell of superstition.

But preferring instead to follow Durkheim and Levy-Bruhl most later European anthropologists broke decisively with the idea of primitive thought as an early if defective kind of science.[7] Anthropologists now claimed primitive thought as "*a wholly different method of conceptual processing*"[8] – as a symbolic logic.[9]

Once primitive thought came to be seen in terms of a logic of symbols and no longer as 'failed science' social anthropologists and scientists went separate ways; there was no longer anything to discuss. Anthropologists largely neglected to ask about the social and cognitive basis of science, preferring instead to concentrate upon sifting vanishing tribal cultures for symbols.

The inductivist continuum was defended seriously only within the small specialist (and mainly North American) sub-discipline of ethnoscience[10], and by the occasional maverick elsewhere.[11] Ethnoscience, concentrating on plant and animal classification among non-literate peoples,[12] dug for global taxonomic principles underlying local knowledge. The search for such global principles offered a challenge to the dominant cultural relativism of modern anthropology.

Spurred on by debate concerning the nature and evolution of human cognitive capacities ethnoscience has experienced something of a revival in recent years.

Scott Atran has been an important fgure in this re-evaluation.[13] The point of Atran's work (as summarised above) is to draw attention to the fact that scientists and anthropologists see science and symbolism as opposed only because they ignore the middle ground. This middle ground comprises common sense.

To understand the relationship between experience and common sense is to understand that science is not 'special creation' but an extension of ordinary human capacities. Common sense, Atran argues, "*lights a world for all to see ... science may excavate and extend that world in wholly novel ways, but science cannot just confute it and render it useless.*"[14] Atran concludes that common sense is an independent but faithful ally of science.

Common sense understanding of the natural world

This section briefly sketches some examples of common sense understanding of the natural world.

All human groups name and group plants and animals. To do so is practical common sense.

Anthropologists learn about naming and grouping plants and animals from their first day in the field. The recently arrived and socially inept anthropologist is unsuited to polite conversation, but is at least fit for work in the fields.

Being able to recognise weeds and distinguish them from crops is one of the most basic qualifications for being invited to join a farm work group.[15] The new recruit has only to match names with plants and more experienced workers can then shout words of advice or warning.

This work of naming and grouping plants and animals can be labelled "indigenous taxonomy". The material has proven a rich field for ethnoscientific investigation. Such taxonomies have to work. Fancifulness is a luxury that cannot easily be afforded. Whether the work done by an indigenous taxonomy is social, technical or practical is a matter of dispute. It may be sensible to assume that common sense taxonomy serves multiple purposes.

All human groups also possess abundant reserves of common sense concerning natural history – i.e. the nature and behaviour of plants, animals and natural phenomena such as soils and rivers.[16] Untutored farmers frequently demonstrate good practical knowledge not only of what crop types suit which soils, but of basic principles of plant selection. They know that not all seed reproduces true to type. A farmer looking for yellow cobs will not select yellow seed from a maize cob of mixed blue and yellow corn. Rice farmers in Sierra Leone know that seed rice from the edge of the farm (where some natural outcrossing has taken place with neighbouring plots) will be less likely to reproduce to type than seed reaped in the middle of the farm.[17] Longley & Richards report that some farmers plant mixed types in the same field hoping to encourage hybridization.[18]

The common sense of indigenous farming is now quite well documented in the scientific literature, being closely linked to the interest and practical concerns of agro-ecologists and agronomists.[19] A more neglected field might be termed "common sense ethology", i.e. common-sense knowledge of animal behaviour.[20]

Anthropologists have written much more copiously about animals as natural symbols than about common-sense natural history.[21] Local hunters may elaborate symbolic themes in fireside tales, but they take very great care not to be wrong about the true behavioural characteristics of the animals they hunt. Their lives may depend on it. Where there is genuine confusion about the true character of an animal – perhaps because its behaviour has been modified by habitat loss or hunting pressure – hunters are often assiduous in monitoring their prey. Puzzled by the chimpanzee – is it some kind of proto-human, capable of evil? – hunters in Sierra Leone sometimes take a common-sense experimental approach. One informant even borrowed a tame chimp from a trader to study its behaviour more closely.[22]

Common sense and nonsense

How do we decide that common sense is not in fact popular nonsense?

It is counter-intuitive for science to group bats and whales with the mammals. But is it popular nonsense to see these creatures as birds and fish in the first place?

Atran answers "no."[23] Dropping bats from the birds and cetaceans from the fish and joining them to the mammals requires special interior knowledge. The common sense understanding makes perfect practical sense. Human beings experience bats and cetaceans as they would birds and fish – bats swoop on the evening breeze like swallows, porpoises rock canoes and tangle nets like sharks. Knowing these things may help avoid capsize, or shock.

Common sense may differ from science, but unlike popular nonsense, it does not jar with what scientists know to be true. Outside the laboratory scientists are quite happy to rely upon common sense. As Atran points out 'tree' is not a valid taxon in botany, but a botanist walking in the forest finds it useful to keep in mind a 'class' of large upstanding plants with hard and often sharp projections. Counter-intuitive purists might otherwise end up with a sore head; oil palm 'trees' are especially unforgiving when banged into unprovoked.

Surveying work on "folk classification" Atran concludes that common sense is no incoherent *mélange* of symbol, speculation and misinformation.[24] Common sense knowledge of nature is both experientially true and convenient in use. If there are gaps in the picture it is because not everything in the biological world can be experienced locally. Problems loom only where experience ends and speculation begins.

The meeting ground of common sense and bioscience

Local environments vary – common sense varies with environment. All human groups have common-sense knowledge of nature, but since the content of common sense varies by locality there is no such single thing as "global" common sense (i.e. common sense covering the entire natural world).

According to the inductivist understanding of the relation between science and common sense, science takes off from local understandings and reaches out for global understanding. Democratic, "bottom-up", conservation seeks to trace out the same trajectory. The challenge for proponents of participatory conservation is to build bridges between local environmental knowledge and global conservation science.

How might such bridges be built, when there is no global common sense? Common sense may be local in content, but ethnoscience stresses the point that a capacity for common sense depends on faculties possessed by humans as a species. In other words, the content of common-sense knowledge may be local but the capacity to know is global. The capacity for common-sense is in this respect not in any obvious sense different from the capacity to know science. But to say there are no cognitive barriers to building a platform to unite common sense and science in the conservation cause is only a starting point. Much institutional (and political) effort will then be required to realise the democratic platform envisaged.

The remainder of this paper considers practical examples of alliance between science and common sense.

Communication between common sense and bioscience

One of the commonest meeting points between common sense and bioscience comes at the outset of a biological project, when field workers in unfamiliar terrain seek local advice about distribution and habits of local flora and fauna. How do scientists and local informants communicate?

The journals of the Swedish botanist Adam Afzelius, working in Sierra Leone in the 1790s, are an interesting source in regard to this question.

For Afzelius common sense was clearly a daily "faithful ally" of his science. The journals are rich in detail about how Afzelius acquired specimens, from whom, in what circumstances, and with what input of local knowledge. On one occasion local knowledge also proved a life-saver. Driven by French bombardment from his house to a hut in the forest, Afzelius was nursed back to health from a near-fatal fever by a passing African "princess" drawing upon a knowledge of forest leaves still today part of the practical medicine chest of every Sierra Leonean women. It is recorded that the "doctor" was so pleased with her patient's recovery that she danced for him before wending her way.

Reading Afzelius' journals raises the question of how the 'Darwins' and 'Man Fridays' of biological exploration managed to communicate. The alacrity with which the great naturalists seemed to attack their work, collecting and exchanging information on plants and animals with locals even on fleeting visits, suggests that biological common sense is relatively unhindered by language difficulties, even where the cultural gap is large.

Afzelius may possibly have conducted his researches in a variant of the West African "*Coast Pidgin*" spoken in the vicinity of Sierra Leone. This was widely used by the motley collection of longer-term white residents on the West African coast, especially those married locally. A creole derived from this coastal pidgin is today the national *lingua franca* of Sierra Leone.

If Afzelius used pidgin to communicate, it may be a point of more than circumstantial significance. Linguists have argued humans have a *"language instinct"* – i.e. a capacity for more-or-less immediate rough-and-ready vocal or sign-based communication, invoked where the parties have no language in common. These pidgins share universal grammatical features even where there has been no prior contact between the contributing parties. This universality can be explained under the assumption that the grammar of pidgins reflects the way in which human brains are organised for language. Nevertheless, pidgins require a social context before any universal capacity for pidgin communication is invoked. Deaf children all use signing to communicate, but only where this signing develops in a group context does the resulting sign language show pidgin-like grammatical features.

Further work will be needed to explore the possibility that the human capacity quickly to "make sense" when communicating about plants and animals is based, at the cognitive level, on mental processing similar to that involved in pidgin communication. But there can be no doubt this is an important research priority. It would open the way to the testing of one of Atran's major claims that better understanding of the way human brains process information about the external world provides the key to understanding common sense as a universal human attainment.

Colonial bioscience and local knowledge of the natural world

The historical record is a rich source of evidence concerning the ways in which colonial bio-science often drew systematically on local common sense knowledge of the environment.

A key point is that colonial bio-scientists were resident 'commoners', not (like Darwin) a visiting 'elite'. Progress in colonial bio-science depended crucially on how well the 'commoners' – typically, hard-working field officers attached to agricultural, forestry or health departments – were able to establish communication with and learn from 'local experts'.

John Ford, a British colonial entomologist, made much of this factor in his magnum opus on the evolution of understanding of the African sleeping sickness problem.[25] Ford was able to challenge elite and metropolitan understandings of the disease because he lived with it for so many decades, drawing on the integrated 'domestic' practical and historical knowledge of local farming and cattle-herding peoples among whom he worked. This was the key to his ecological interpretation of trypanosomiasis, at odds with the 'reductionist' understandings issuing from well-found disease-research laboratories in Europe.

Another representative 'resident commoner' of colonial bio- science was the late Frederick Deighton, who spent his entire professional career (1926-57) as government plant pathologist in Sierra Leone. Within ten years of taking up his post Deighton had trekked to every chiefdom in Sierra Leone, laying the foundations for an ethnobotanical dictionary listing names for a major part of the national flora in local languages. His death finally interrupted a huge project to classify the Sierra Leonean funguses. Thirty-five years after retirement he still maintained contacts with 'local experts' trained to care for herbarium collections first established in the 1930s. Specimens misplaced in London were still safely under lock and key in 1990 in a back-up collection established at Njala, a small rural agricultural college in central Sierra Leone.

Tracking the career and contacts of a scientist like Deighton we begin to understand the role of an unsung army of skilled local assistants – trackers, rangers, tree spotters, line cutters, guides, interpreters, herbarium curators and daily labourers, who piloted colonial bio-science researchers through unknown terrain, helping locate, protect and keep in order the basic materials without which generalised scientific conclusions would have been unthinkable. Often poorly educated by formal standards, these 'subalterns' made an unquantifiably large contribution to, for example, tropical forest ecology and the Green Revolution. Only occasionally is this contribution formally acknowledged and brought to the attention of a wider audience.[26] Colonial bioscience, we could say, was an edifice erected upon foundations in which common sense, folk knowledge and science were thoroughly mixed.

Experimentation in common sense and science

Problem solving by experimentation is another area where convergence between common sense and science is readily apparent.

Taken by-and-large there is no doubt that the dominant human approach to problem solving is social learning. Social learning – copy, be taught, or ask advice – is cost-effective[27] where environmental conditions are stable. Much of common sense comprises the social skills (deference, humility, a willingness to practice hard and be corrected) to make effective use of the knowledge of those with the greatest experience of life. But social learning does not always work. Circumstances change or the problem may be unprecedented. Maybe there is nothing for it but to experiment.

It was once thought that 'indigenous groups' did little experimentation. They were even called "traditional societies". More recent ethnography has altered this picture. It is clear that experimentation is a feature of many supposedly "traditional" communities.[28] The reason seems clear. Earlier accounts greatly over-estimated the stability of the environmental conditions under which these communities lived.

European overseas expansion, from the 15th century onwards, greatly altered the biological conditions of existence for the majority of the native peoples of Asia, Africa and the Americas.

Where native groups were not exterminated or forced on to reserves they could sometimes benefit, in areas of successful European colonization, from introduced technologies. The picture is complex, and there was often two-way learning – subtle hybridizations – between colonizing and subject peoples. An important North American instance is discussed in Jordan & Kaups'[29] study of pioneer agriculture in the 17th century Upper Delaware Valley. But essentially adaptation to changed environmental circumstances was achieved through adoption of innovation (a form of social learning), not through experimentation.

Tropical Africa, however, presents a very different picture. There were few viable settler enclaves. Yet the environment was greatly changed by rampant "primitive accumulation" from the period of the slave trade onwards. Entire regions were depopulated by slave raiding. New diseases, or epidemic outbreaks of hitherto localised forest pathogens, spread along long-distance trade routes, with devastating consequences for human fertility, cattle economies and management of intensive agriculture. Forest ecologies were disrupted by the search for wild rubber, lumber, ivory and precious minerals. Young men were press-ganged into carrying loads long-distances, or were later recruited for mines in South Africa, forcing women to shoulder the burdens of agricultural subsistence alone.

Nineteenth century equatorial Africa as described, for example, by Ford[30] is a world of environmental dislocation and flux. There were few if any ready-made answers from outside to deal with these problems. That is the central point of Ford's review of colonial attempts to "solve" the Sleeping Sickness problem; outside intervention made matters worse. Tropical Africans had little option but to cope with these changes on their own. They became great experimenters as a result.

Social learning is in most societies strongly institutionalized. Even where there are no schools it is taught through, say, respect for the authority of elders. Experimentation is regularly perceived as a threat to the major processes through which wisdom and experience are conserved.

The problem for environmentally unstable 19th century Equatorial Africa was how to create space for experimentation without undermining social learning. Or put in more conflictual terms, the problem was how to win a space where experimentation was tolerated.[31]

Experimentation, as a tool of common sense, takes many forms. Social experimentation might require its own cultural 'space'. One way of viewing so-called "secret societies" is to see them as guilds engaged in 'domesticating' wild and unpredictable spirit forces associated with the forest. Such guilds abrogated to themselves the dangerous business of "experimenting" in the "laboratory" of spirit forces.

More mundane examples are to be found in the field of crop science. Mende-speaking villages, in Sierra Leone, for example, recognise experimentation with rice planting material as a legitimate, and praiseworthy, activity. Everyone needs good seed. Useful varieties are quickly multiplied and given, begged or stolen. This type of experimentation was rarely perceived as a threat to authority.

Mende culture recognises that the agricultural environment changes, and that farmers need 'space' to tinker with that environment. Some of the best agricultural experimenters, in my experience, are community outsiders – migrant 'strangers'. Strangers need to experiment to cope with a new environment.

Different words in the Mende language discriminate between 'tests' or 'experiments' as trials of strength between persons (as in court cases) and 'experiments with nature' as seen on the farm. These are both events with uncertain outcomes. The former concerns society, the latter, nature. Hacking suggests that the ability to distinguish and then to recombine two distinct notions of probability – 'aleatoric', (i.e. dice- throwing) probability, referring to variation in nature, and 'epistemic' probability, referring to degrees of belief, was foundational to the emergence of the modern theory of statistics upon which experimental science now rests.[32] It is intriguing to find the same distinction apparent in Mende common sense.

In other cases it was authority that undertook the experiment. Ford (1971) describes one dramatic instance, in which a leader of the Mfecane expansion through early 19th century southern Africa, moving his people into tsetse infected country on the present-day borderlands of Zimbabwe and Mozambique, insisted on a policy of settlement concentration as a practical measure to eliminate the sleeping-sickness threat.[33]

There seems no mystery about how early colonial science in Africa so quickly found and trained a cadre of capable scientific assistants. Under nineteenth century conditions of environmental instability common sense in tropical Africa had already developed a strong experimental turn.

Conclusion

Enough has been said to illustrate Atran's claim that there is nothing in common sense, separated from symbolism and popular nonsense, hostile to science.

Indeed there are areas of common sense – e.g. problem-solving experimentation – that overlap the terrain of the scientist. Common sense provides a potential platform, I conclude, for an informed participation in debate about science-guided conservation plans. There are various ways in which such platforms might be built – e.g. through the medium of "consensus conferences", where representative citizens arrive at a common-sense verdict after gaining access to a wide cross-section of expert scientific opinion, or through attempts to introduce a problem-solving approach into the school science curriculum (see, for example, the interesting discussion by Samuel Mokuwa of how the problem-solving curriculum approach might help lay the foundations for greater citizen involvement in plant genetic resource management issues in Sierra Leone[34]). But bringing common sense and science together in the interests of better management of bioresources, clearly, will vary with locality and circumstances.

Two points arise. First, as illustrated by the discussion of African experimentation, some common senses are more finely attuned to environmental change than others. Further analysis is now needed to work out how different common senses vary in their affinity for science. Second, we should not lose sight of the fact that common sense is local knowledge, but that conservation increasingly requires global solutions. Democratic participation in global conservation implies a very complex process of networking and coalition building. Further hard work is needed to understand how to build effective coalitions out of non-coincident bodies of local common sense. Perhaps the single biggest challenge for those of us interested in the democratization of conservation will be to establish a conceptual framework within which local knowledges will be seen and respected as instances of a global human capacity for common sense. Some talk, in this context, of hybridization or creolization, but maybe the true word for the process that unites all our common senses, without hegemony or hierarchy, authority or cant, is 'science'. The task is to move from 'indigenous knowledge' to 'people's science'.

Notes

1   Cited in Atran, S.: *Cognitive foundations of natural history: towards and anthropology of science*, Cambridge: Cambridge University Press 1990, p. 275.
2   *Op. cit.*
3   *Op. cit.*
4   Wolpert, L.: *The unnatural nature of science*, London: Faber 1992.
5   Atran, 1990, *op. cit.*, p. 247.
6   *Op. cit.*, p. 265.
7   Tambiah, J. S.: *Magic, science, religion and the scope of rationality*, Cambridge: Cambridge University Press 1990.

8   Atran, 1990 *op. cit.*
9   Douglas, M.: *Purity and danger,* London: Routledge 1966.
10 Fowler, C. S.: "Ethnoecology", in: Hardesty. D. (ed.): *Ecological Anthropology,* New York: Wiley 1977, ch.12. & Conklin, H.: *Folk classification: a topically-arranged bibliography of contemporary and background references through 1971.* Department of Anthropology, Yale University 1980.
11 Cf. Horton, R.: "African traditional thought and Western science," *Africa,* Vol. 37, 1967, pp. 50-71 and 159-187.
12 E.g. Berlin, B., Breedlove, D., and Raven, P.: *Principles of Tzeltal plant classification,* New York: Academic Press 1974.
13 Atran 1990, *op. cit.*
14 Atran 1990, *op. cit.,* p. 269.
15 Richards, P.: *Coping with hunger: hazard and experiment in a West African farming system,* London: Allen & Unwin 1986.
16 Richards, P.: "Natural symbols and natural history: chimpanzees, elephants and experiments in Mende thought," in: K. Milton, (ed.): *Environmentalism: the view from anthropology,* ASA Monograph 32, London: Routledge 1993.
17 Richards, P.: "Culture and community values in the selection and maintenance of African rice," in: S. Brush and D. Stabinsky (eds.): *Intellectual property and indigenous knowledge,* Covelo CA: Island Press 1996.
18 Longley, C. and Richards P.: "Farmer innovation and local knowledge in Sierra Leone," in: K. Amanor, W. de Boef, A. Bebbington, & K. Wellard, (eds.): *Cultivating knowledge,* London: Intermediate Technology Press 1993.
19 Richards, P.: *Indigenous agricultural revolution: ecology and food production in West Africa,* London: Hutchinson 1985.
20 Richards 1993, *op. cit.*
21 Douglas 1966 and Richards 1993, *op. cit.*
22 Richards 1993, *op. cit.*
23 Atran 1990, *op. cit.*
24 Atran 1990, *op. cit.*
25 Ford, J.: *The Role of Trypanosomiasis in African Ecology,* Oxford: Oxford University Press 1971; Richards 1985, *op. cit.*
26 Cf. Barnish, G. & Samai, S. K.: *Some medicinal plant recipes of the Mende, Sierra Leone.* Bo: Medical research Council/SLADEA 1992.
27 Collier, P.: "Africa and the study of economics," in: R. Bates, V. Mudimbe & J. O'Barr, (eds.): *Africa and the disciplines.* Chicago: University of Chicago Press 1992.
28 Richards 1986, *op. cit.* & Guyer, J. and Belinga Eno, S. M.: "Wealth in people as wealth in knowledge: accummulation and composition in Equatorial Africa," *Journal of African History* Vol. 36, 1995, pp. 91-120.
29 Jordan, T.G. and Kaups, M.: *The American backwoods frontier: an ethnic and ecological interpretation,* Baltimore: John Hopkins University, 1989.
30 Ford 1971, *op. cit.*
31 Guyer and Eno Belinga 1995, *op. cit.*
32 Hacking, I.: *The emergence of probability,* Cambridge: Cambridge University Press 1975.
33 Ford 1971, *op. cit.*
34 Mokuwa, S.: *Rice biodiversity education: a problem-based curriculum innovation for post-war recovery,* MSc Thesis, MAKS, Wageningen Agricultural University, NL 1997.

Utilizing Amazonian indigenous knowledge in the conservation of biodiversity:

Can Kayapó management strategies be equitably utilized and applied?

Darrell A. Posey

Introduction: recognizing the inextricable link

The Convention on Biological Diversity (CBD), opened for signature during the 1992 Earth Summit and ratified by over 160 countries, has become the major global initiative to stop the loss of biological and cultural diversity. Indigenous peoples and local communities are recognised as playing important roles in *in situ* conservation by having sustainably managed the natural resources of fragile, biologically rich ecosystems for millennia. This is specifically noted in the Preamble of the CBD, which acknowledges the *"close and traditional dependence of many indigenous and local communities embodying traditional lifestyles on biological resources, and the desirability of sharing equitably benefits arising from the use of traditional knowledge, innovations and practices relevant to the conservation of biological diversity and the sustainable use of its components."*

The section that is most obviously and directly related to indigenous peoples is Article 8j. It states that each Contracting Party should *"Subject to its national legislation, respect, preserve and maintain knowledge, innovations and practices of indigenous and local communities embodying traditional lifestyles relevant for the conservation and sustainable use of biological diversity and promote the wider application with the approval and involvement of the holders of such knowledge, innovations and practices and encourage the equitable sharing of the benefits arising from the utilisation of such knowledge, innovations and practices."*

Indigenous peoples themselves frequently emphasise that conserving biological diversity in areas such as forests requires respect and recognition of their rights, as in the 1992 Indigenous Peoples' Earth Charter:[1] *"Recognising indigenous peoples' harmonious relationship with Nature, indigenous sustainable development strategies and cultural values must be respected as distinct and vital sources of knowledge."* (Clause 67)

Similarly, the Final Statement from the 1995 Consultation on Indigenous Peoples' Knowledge and Intellectual Property Rights in Suva, Fiji[2] and the 1992 Charter of the Indigenous-Tribal Peoples of the Tropical Forests[3] assert that indigenous guardianship of ecosystems is the best method of conserving biological diversity and indigenous knowledge. In the specific context of forests, the Indigenous-Tribal Peoples' Charter states: *"All policies towards the forests must be based on a respect for cultural diversity, for a promotion of indigenous models of living, and an understanding that our peoples have developed ways of life closely attuned to our environment."* (Article 5)

Thus, the interdependence of cultural and biological diversity is clearly recognised by both the international community and indigenous peoples, as is respect for the rights of indigenous peoples as a *sine qua non* for sustainable use and conservation of forests and their biodiversity. This also requires respect for local values and equitable use of traditional knowledge as a basis for sustainable management.

Local knowledge and sustainable management

The CBD language of *"traditional knowledge, innovations and practices"* of *"local communities embodying traditional lifestyles"* is usually referred to by scientists as Traditional Ecological Knowledge (TEK).[4] TEK is far more than a simple compilation of facts. It is the basis for local-level decision-making in areas of contemporary life, such as natural resource management, nutrition, food preparation, health, education, and community and social organisation.[5] TEK is holistic, inherently dynamic, constantly evolving through experimentation and innovation, fresh insight, and external stimuli.[6]

According to the Four Directions Council:[7] *"What is 'traditional' about traditional knowledge is not its antiquity, but the way it is acquired and used. In other words, the social process of learning and sharing knowledge, which is unique to each indigenous culture, lies at the very heart of its 'traditionality'. Much of this knowledge is actually quite new, but it has a social meaning, and legal character, entirely unlike the knowledge indigenous peoples acquire from settlers and industrialised societies. This is why we believe that protecting indigenous knowledge necessarily involves the recognition of each peoples' own laws and their own processes of discovery and teaching."*

The goal of linking indigenous systems and sustainability should be to harness the totality, rather than the components, of TEK systems in sustainability strategies, so that the holistic quality of indigenous management dominates. Policy-makers need to be aware, however, that indigenous peoples frequently integrates forest and agricultural management systems so that categories of 'forester', 'hunter and gatherer' and 'farmer' cannot be considered as discrete categories, but represent continua of traditional livelihood activities.

The continuum between agriculture and forestry is filled by the vast number of plant and animal species that are *neither* agricultural domesticates *nor* timber species, but nonetheless provide most of the needs of local communities. Many of these species have been genetically selected, planted, and transplanted to enhance and modify local ecosystems. These are sometimes known as 'semi-domesticates' or 'human modified species'[8,] although *Non-Domesticated Resources* (NDRs) is my preferred term. NDRs have systematically been undervalued and overlooked by scientists, yet provide a vast treasury of useful species for food, medicines, shelter, building materials, dyes, colourings, repellents, fertilizers, and pesticides.[9]

Many so-called 'natural' landscapes are in fact *cultural* or *anthropogenic landscapes*, either of cultural significance or modified by human activity. Failure to understand the human impact on 'wild' landscapes has blinded outsiders to NDRs and the traditional management practices that determine their use and conservation.[10] Indigenous peoples and a growing number of scientists find unacceptable the assumption that just because landscapes and species appear to outsiders to be 'natural', they are 'wild' and – therefore – unowned.

According to a Resolution from the 1995 Ecopolitics IX Conference in Darwin, Australia: "*The term 'wilderness' as it is popularly used, and related concepts as 'wild resources', 'wild foods', etc., [are not acceptable]. These terms have connotations of terra nullius [empty land] and, as such, all concerned people and organisations should look for alternative terminology which does not exclude Indigenous history and meaning.*"[11]

Scientific studies that document the anthropogenic and cultural aspects of landscapes, therefore, can overturn the 'empty land' concept and begin to recognise the role of indigenous and local peoples in *in situ* conservation. Theoretically, this will in turn strengthen claims by communities to ownership of and rights over land, knowledge, and genetic resources.

Cultural landscapes and Kayapó resource management

Traditionally, Amazonian Indians have been thought of as merely exploiters of their environments – not as conservers, manipulators, and managers of natural resources.[12] Researchers are finding, however, that presumed 'natural' ecological systems may, in fact, be products of human manipulation.[13] Likewise, old agricultural fallows reflect genetic selection and human enhanced species diversity.[14]

The Kayapó Indians of the Middle Xingu Valley, Brazil, provide a good example of how scientific assumptions of 'natural' landscapes have hidden the complexity and potential of local management practices to modify ecosystems. The modern Kayapó population is still under 5,000, but pre-contact populations were many times larger and presumably had even greater impacts on the vast region they exploited. They live in an ecologically diverse region that comprises nearly 4 million hectares of *reserva indigena* in the States of Para and Mato Grosso. Ethnohistorical research with the Kayapó Indians shows that contact with European diseases came via trade routes and preceded face-to-face contact with colonisers. Epidemics led to intra-group fighting, fission, and dispersal of sub-groups that carried with them seeds and cuttings to propagate their foods, medicines, and other resources.[15]

A form of 'nomadic agriculture' developed based upon the exploitation of non-domesticated resources (NDRs) intentionally concentrated in human-modified environments near trail sides, abandoned villages and at camp sites. Agricultural practices also spread, along with techniques for management of old fields, to enhance availability of wildlife and useful plants. During times of warfare, the Kayapó could abandon their agricultural plots and survive on non-domesticated species concentrated at trail sides, former village sites, forest openings, and ancient fields.

Agricultural plots were engineered to develop into productive agroforestry reserves dominated by NDR species, thereby allowing the Kayapó to oscillate between (or blend together) agriculture and gathering. Such patterns appear to have been widespread in the lowland tropics and defy the traditional dichotomies of wild vs. domesticated species, hunter-gathers vs. agriculturalists, and agriculture vs. agroforestry. Even today, over 76% of the useful plant species collected to date are *not* 'domesticated', nor can they be considered 'wild'.[16] (I suspect that as a more complete floral inventory is completed, this percentage will approach 98%).

Nowhere is this more evident than in the formation of "islands of forest", or *apêtê*, in the *campo-cerrado* (savanna). The Kayapó initiate and simulate the formation of forest patches through the careful manipulation of micro-environmental factors, knowledge of soil and plant characteristics, and intentional concentration of useful species into limited plots [See Figure 1.]. Although most *apêtê* are small (under 10ha), elders reported plant varieties in a 1ha plot as having been introduced by villagers from an area the size of Western Europe.[17]

The principle elements of Kayapó management have been previously described in some detail,[18] and include:

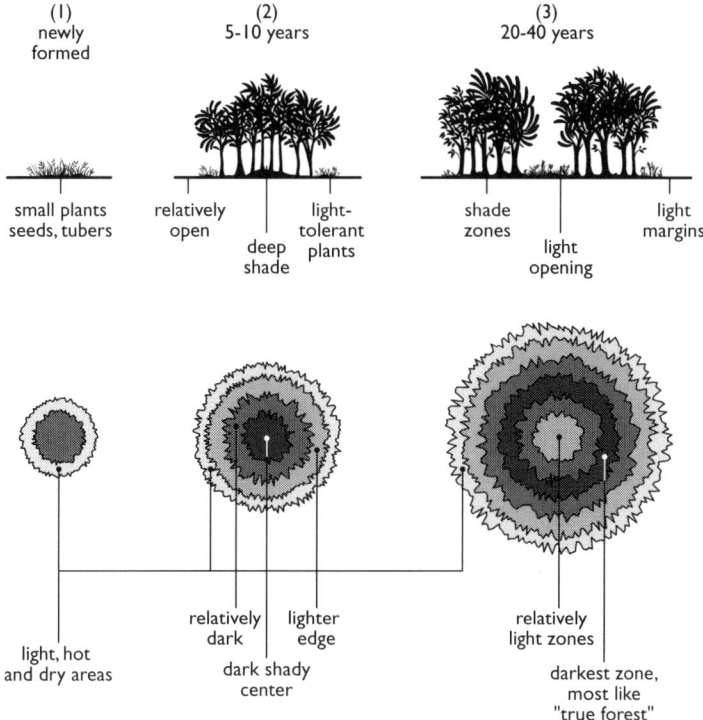

Figure 1. Cycle of anthropogenic 'islands of forest' (apêtê) formation in the campo-cerrado

1. overlapping and interrelated ecological categories that form continuua
2. modification of 'natural ecosystems' to create ecotones
3. emphasis on long-term ecotone utilisation (*chronological ecotones*)
4. concentration on non-domesticated resources
5. transfer of useful plant varieties between similar ecological zones
6. integration of agricultural with forest management cycles

Several options are possible for representing indigenous resource management models. I believe the most inclusive and descriptive representation of the Kayapó system places savannah or grass lands (*kapôt*) at one end of a continuum as the 'focal type' (example that most typifies the category) and forests (*bà*) at the other (opposite focal type). *Kapôt* types with more forest elements would be represented to the right of the diagram, while *bà* types that are more open and with grassy elements would lie on the continuum diagram to the left, or toward the savanna pole.

This would put *apêtê* at the conceptual centre of the continuum, since forest elements are introduced into the savanna to produce these anthropogenic zones.

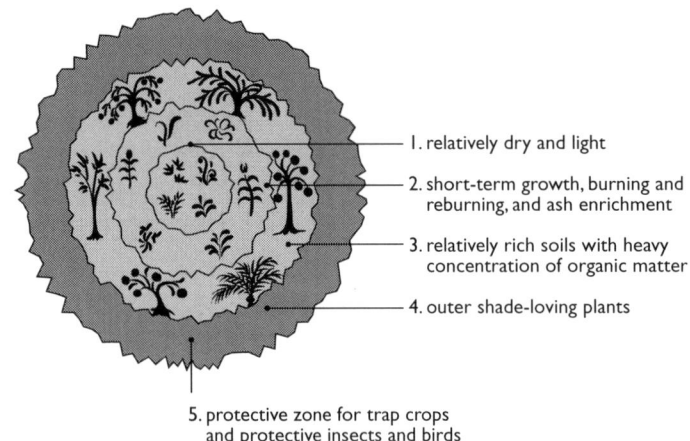

Figure 2. Idealised gardens (puru), showing microzones

Agricultural plots (*puru*) also lie conceptually near the centre of the continuum, because sun-tolerant vegetation is introduced into managed forest openings [Figure 2]. *Apêtê* can be thought of as the conceptual inverse of *puru*: the former concentrates resources in the forest using sun-tolerant species, while the other does the same in the savanna using forest species.

Even though ecological types like high forest (*bà tyk*) or transitional forest (*bà kamrek*) are securely located at the forest pole, they are not uniform in their composition. All forests have edges (*kà*), margins (*kôt*), and openings caused by fallen – or felled – trees (*bà krê-ti*) that provide zones of transition between different conceptual zone. Thus, a plant that likes the margins of a high forest might also grow well at the margin of a field *(puru-kà,* or *puru-kôt)* or in an *apêtê*. A plant that likes light gaps provided by forest openings might also like forest edges (*bà-kà,* or *bà-kôt*) or old fields (*puru-tum* or *ibe-tum*). Plants from open forest types or forest edges can predictably proliferate along edges of trails or thicker zones of *apêtê*. Using this logic, the Kayapó can transfer biogenetic materials between matching micro-zones so that ecological types are interrelated by their similarities rather than isolated by their differences. These interfaces can be considered ecotones, which become the uniting elements of the overall system.

There is another interesting dimension to the model that appears when looking diachronically (temporally or historically) across the system. Agricultural clearings are initially planted with rapidly growing domesticates, but almost immediately thereafter are managed for secondary forest and NDR species.

This management depends upon planting and transplanting, removal of some varieties, allowing others to grow, encouraging some with fertilizer and ash, and preparing and working the soils to favour useful species.

Management aims to provide long term supplies of building materials, ceremonial objects, medicinals and other useful products, as well as food for humans and animals. The old fields (*puru tum*) are at least as useful to the Kayapó as agricultural plots or mature forest.[19] A high percentage (an initial estimate is 85%,) of plants in this transition have single or multiple uses. When the secondary forest grows too high to provide undergrowth as food for animals (and hunting becomes difficult), then the large trees are felled to create more hospitable conditions for management and/or reinitation of the agricultural cycle. Likewise, *apêtê* are managed to maximise useful species in all stages of the forest succession. When their centres become dark and unproductive, openings (*irã*) are created that allow light to again penetrate the forest and reinitiate a new cycle.

The Kayapó resource management system is, therefore, based on the conservation and use of transitional forests in which agriculture is only a useful (albeit critical) phase in the long-term process. *Apêtê* exhibit parallel transitional sequences in the campo-cerrado and depend almost exclusively on non-domesticated resources. The degree to which genetic material are transferred between similar micro-zones of different ecological types points to how the Kayapó exploit ecotones that host the highest diversity of plants. Management over time can be thought of as management of *chronological ecotones*, since management cycles aim to maintain the maximum amount of diversity and the greatest number of ecotones.

The Kayapó model illustrates how previously assumed 'natural' ecosystems in Amazonia have been consciously modified by indigenous residents through time. The degree to which this has taken place has yet to be quantified, but Kayapó "forest islands" data show concentrations of plant varieties from a vast geographic area. This case underlines the necessity for historical studies to understand the long-term effects of management of cultural and anthropogenic landscapes. Above all, it exposes the inadequacies of our scientific, educational, and political institutions that separate agriculture from forestry and ignore the importance of non-domesticated resources.

Can Kayapó TEK be put to "wider use and application"?

The CBD recognizes various ways in which traditional knowledge, innovations, and practices can be employed to enhance *in situ* biodiversity conservation. The most fundamental is recognition of community and cultural rights that allow indigenous peoples like the Kayapó to make their own decisions over the future of their land and resources. A second way is through community controlled conservation and full participation of indigenous and local peoples in all aspects of planning and implementation of conservation and development projects.[20]

These strategies are founded upon respect for and preservation of the holistic nature and diversity of indigenous natural resource management systems. But most interest in TEK is not concerned with *in situ* conservation, but rather the use and application of knowledge and genetic resources for the development of natural products.

Industry has discovered the economic potential for new products 'mined' from the biodiversity rich areas using TEK to provide 'leads' or short cuts in product discovery.[21] This is known as 'bioprospecting' and high expectations of profits have led to frenzied collecting and commercial activity on Kayapó lands and elsewhere in Amazonia.

It is difficult to estimate the commercial promise of biodiversity prospecting. For medicines alone, the 1985 market value of plant-based medicines sold in developed countries was estimated at a total of $43 billion.[22] This frequently cited estimate is unreliable, but whatever the true amount might be, only a minuscule proportion of profits have been returned to Indigenous peoples from whom much of the original knowledge came. I have estimated less than 0.001% of global sales.[23]

Companies that produce seeds and agrochemicals benefit substantially from the free flow of germplasm from Indigenous lands. The market value of the seed germplasm utilising traditional landraces is estimated by Rural Advancement Fundation International at $50 billion per year in the US alone.[24] Consequently, indigenous peoples are providing subsidies to a modern agricultural system that barely recognises their contributions. Similar situations exist with timber and non-timber forest products, as well as other natural product markets, such as personal care, foods, industrial oils, essences, pesticides, and preservatives.

Biodiversity prospectors assume that organisms and ecosystems are 'wild' and, therefore, part of 'the common heritage of humankind'. Even when critical information, or even processed materials, are provided by Indigenous peoples, it is the company that makes the protectable or patentable 'invention' and acquires the financial gains. Indigenous peoples see this situation as being parallel to the Europeans' "discovery" of the New World and are understandably weary of the process. In the Pacific region, indigenous peoples have called for a total *moratorium* on bioprospecting, a move that has been echoed in Ecuador and is under consideration in indigenous territories around the world.

Scientists and scientific institutions are affected by this situation as they become involved – actively or passively – with the private sector. Plant, animal, and cultural material collected with public funds for scientific, non-profit purposes are now open for commercial exploitation. Research, even in universities and museums, is increasingly funded by corporations, raising questions of who controls the resulting data. 'Purely scientific' data banks have become the "mines" for 'biodiversity prospecting'. Publishing of information, traditionally the hallmark of academic success, has become a superhighway for transporting restricted (or even sacred) information into the unprotectable 'public domain'.

Many scientific institutions and professional societies are responding to this dilemma by developing their own codes of conduct and standards of practice. These generally follow the well-established principles in international law and customary practice that include *inter alia*: recognition of indigenous land, territorial and intellectual property rights; support of indigenous self-determination; collective or community rights; full disclosure of intent; prior informed consent; and, veto over research and development projects. A good profile of these principles can be found in the Findings and Recommendations of a Workshop on "Indigenous Peoples and Traditional Resource Rights" of the Oxford Green College Centre for Environmental Policy and Understanding [APPENDIX 1].[25]

It remains to be seen if scientists can aptly apply the TEK they acquire from indigenous and local peoples like the Kayapó to develop alternative models for sustainable forest management. Adequate historical studies of anthropogenic landscapes, recognition of micro-ecological zones and ecotones, development of non-domesticated resource based systems, and exploitation of the forest-agriculture continuum require major shifts in the way science and scientific institutions are organised and funded. And, even with the growing international acceptance of the CBD, enactment of adequate national legislation, and development of professional Codes and Standards of Conduct, there are no guarantees that rates of loss of biological and cultural diversity will slow.

A sincere quest for conservation of diversity, however, provides an unprecedented opportunity for collaboration between peoples and across cultures, and disciplines. Thus, the global environmental crisis can be seen as a unique chance to re-discover and celebrate the co-evolution of humanity with nature, while developing new mechanisms that thrive upon equity and equality–not only of cultures, but of all creatures. The first step is learning from indigenous peoples like the Kayapó and working together to transform conflict into collaboration.

Appendix 1

Findings and Recommendations from the Green College Centre for Environmental Policy and Understanding Workshop: 'Indigenous Peoples and Traditional Resource Rights'. Oxford, 28 June 1995

Finding 1
Indigenous rights are based on concepts of self-determination as defined in relevant declarations. These should guide science, research and development policy as well as efforts to protect traditional resources and intellectual property rights (IPRs). They include:
- territorial and resource rights;
- respect for cultural differences and Indigenous Peoples' own institutions and efforts;
- prior informed consent;
- veto power over research and development projects;

Recommendations
That governmental and non-governmental institutions:
- follow principles already established in indigenous rights documents;
- support, disseminate and integrate these principles into policy guidelines and operations.

Finding 2
Traditional and Indigenous Peoples:
- well express concerns around the world about loss of local autonomy and control, erosion of common resources, and destruction of biological and cultural diversity;
- have inadequate opportunities for dialogue with institutional representatives;
- are under-represented at all levels of governmental and non-governmental decision-making.

Recommendations
That scientists, government and non-government representatives; UN agencies; government departments; scientific and professional institutions:
- recognize and value indigenous knowledge as a basis for new models of development and environmental conservation;

- establish means to facilitate dialogue and form alliances with indigenous leaders;
- strengthen and support local institutions;
- involve Indigenous Peoples in planning and executing projects and policies affecting them and the environments in which they live, and let their knowledge guide all levels of decision-making;
- ensure transparency in all negotiations of research, results, data management, and benefit-sharing;
- establish centres and programmes to guide and facilitate this process.

Finding 3
Modifications of existing practice are necessary to meet the concerns of Indigenous Peoples.

Recommendations
- ensure *in situ* programmes strengthen local livelihoods;.
- make community-controlled research[26] standard practice;
- give local communities prior informed consent and right of veto regarding projects taking place on their lands or territories or that affect them;
- that determination of the *common good* should reflect indigenous and traditional values.

Finding 4
Research and scientific research organizations do not have adequate operational guidelines to reflect the principles of the Convention on Biological Diversity and indigenous rights.

Recommendations
Form a consortium of institutions to:
- establish codes and standards for conduct and policies to reflect indigenous rights and the Convention on Biological Diversity;
- identify gaps between policies and practices, and correct these deficiencies;
- ensure that scientists, government officials, and non-government representatives are properly informed of indigenous rights and views.

Finding 5
Existing IPR instruments are inadequate and new mechanisms must be developed.

Recommendations
- to pursue the 'bundles of rights' approach, to develop Traditional Resource Rights and to look into other legal systems;
- to investigate other ways of protecting intellectual, cultural and scientific resources, including customary practice;
- to observe a moratorium on 'biodiversity prospecting' unless and until adequate and effective mechanisms for protection and compensation have been established.

Finding 6
Institutions may not be able to ensure rights are respected in the countries where Indigenous Peoples reside, but guidelines for institutions can define partners and funding priorities that will affect recognition of indigenous rights.

Recommendations
that as criteria for collaboration:
- indigenous rights, including intellectual property rights are recognized;
- indigenous rights are guaranteed in countries of activity;
- mechanisms are provided to ensure community decision-making, traditional resource rights protection, and benefit-sharing.

Finding 7
Concerns about biosafety are intricately related with concerns about IPRs and TRRs, as release of genetically modified organisms can affect the well-being and livelihoods of local communities.

Recommendations
- to include local communities in the monitoring and evaluation of genetically modified organisms;
- for institutions to exercise the 'precautionary principle' in releasing modified organisms into the environment;
- to look into the concept of 'Life Patent-free Zones' for indigenous lands.

Notes

1. In Posey, D.A. and G. Dutfield: *Beyond Intellectual Property: Toward Traditional Resource Rights For Indigenous and Local Communities*, Ottawa: International Development Research Centre 1996.
2. Pacific Concerns Resource Centre: *Proceedings of the Indigenous Peoples' Knowledge and Intellectual Property Rights Consultation*, Suva: Pacific Concerns Resource Centre 1995.
3. International Alliance of the Indigenous-Tribal Peoples of the Tropical Forests: *Charter of the Indigenous-Tribal Peoples of the Tropical Forests*, Penang: IAI-TPTF 1992.
4. Defined by Gadgil *et al.* as 'a cumulative body of knowledge and beliefs handed down through generations by cultural transmission about the relationship of living beings, (including humans) with one another and with their environment'. Gadgil, M., F. Berkes and C. Folke: 'Indigenous Knowledge for Biodiversity Conservation', *Ambio* vol. 22 1993, pp. 151-156. For other definitions, see Posey and Dutfield *op.cit.*.
5. Warren, D.M., L.J. Slikkerveer and D. Brokensha (eds.): *The Cultural Dimension of Development: Indigenous Knowledge Systems*, London: Intermediate Technology Publications 1995.
6. Although the word 'traditional' is often used to imply 'antithetic to change', recent discussions indicate a shift towards interpreting 'tradition' as a filter through which innovation occurs (e.g. Hunn, E.: 'What Is Traditional Knowledge?', in: Hunn, E. (ed.): *Ecologies For the 21st Century*, Canberra: Australian National University Press 1994, pp.13-15. Pereira, W. and A.K. Gupta: A Dialogue on Indigenous Knowledge. *Honey Bee*, vol. 4 1993, pp. 6-10. Vijayalakshmi, K.: Conserving People's Agricultural Knowledge, in: V. Shiva (ed.): *Biodiversity Conservation: Whose Resources? Whose Knowledge?*, New York: Indian National Trust for Art and Cultural Heritage 1994, pp. 58-72). Innovation is, therefore, a major part of tradition in indigenous and other traditional societies.
7. Four Directions Council: *Forests, Indigenous Peoples and Biodiversity: Contribution of the Four Directions Council*. Draft Paper Submitted to the Secretariat of the Convention on Biological Diversity 1996.
8. Posey, D.A.: 'Importance of Semi-Domesticated Species in Post-Contact Amazonia: Effects of Kayapó Indian Dispersal On Flora and Fauna', in: C.M. Hladik, H. Pagezy, O. Linares, A. Hladik, A. Semple and M. Hadley (eds.): *Food and Nutrition in the Tropical Forest: Biocultural Interactions*, Man and the Biosphere Series (Volume 15). Paris: UNESCO and Parthenon Press 1993, pp. 63-72.
9. Friends of the Earth: *The Rainforest Harvest: Sustainable Strategies For Saving the Tropical Forests. Proceedings of an International Conference Held at the Royal Geographical Society*, London: Friends of the Earth 1992. Kvist, L.P., M.K. Andersen, M. Hellesoe and J.K. Vanclay: 'Estimating use-values and relative importance of Amazonian flood plain trees and forests to local inhabitants',*Commonwealth Forestry Review*, vol. 74 1995, pp. 293-300. Plotkin, M. and L. Famolare (eds.): *Sustainable Harvest and Marketing of Rainforest Products*, Washington DC: Conservation International and Island Press 1992.
10. Gomez-Pompa, A. and A. Kaus.: 'Taming the Wilderness Myth', *BioScience*, vol. 42 1992, pp. 271-279.
11. Northern Land Council: *Ecopolitics IX: Perspectives on Indigenous Peoples' Management of Environment Resources*, Darwin: Northern Land Council 1996.

12 See e.g. Meggers, B.J.: *Amazonia: Man and Culture in a Counterfeit Paradise*. Revised Edition. Washington and London: Smithsonian Institution Press 1996.
13 Alcorn, J.B.: 'Huastec Noncrop Resource Management: Implications for Prehistoric Rain Forest Management', *Human Ecology* vol. 9, 1981, pp. 395–417. Alcorn, J.B.: 'Process as Resource: the Traditional Agricultural Ideology of Bora and Huastec Resource Management and its Implication for Research', in: D.A. Posey and W. Balée (eds.) *Resource Management in Amazonia: Indigenous and Folk Strategies. Advances in Economic Botany*, vol. 7, 1989, pp.63–77. Anderson, A.B. and D.A. Posey: 'Manejo do Cerrado Pelos Indios Kayapó', *Boletim do Museu Paraense Emilio Goeldi, Botânica*, vol. 2, 1985, pp. 77–98. Balée, W.: 'Cultura na Vegetação da Amazônia', *Boletim do Museu Paraense Emilio Goeldi*, vol. 6, 1989a, pp. 95–100 (Coleção Eduardo Galvão). Balée, W.L. and A. Gély: Managed Forest Succession in Amazonia: the Ka'apor Case, in: D.A. Posey and W. Balée (eds.) *Resource Management in Amazonia: Indigenous and Folk Strategies. Advances in Economic Botany*, vol. 7, 1989, pp. 129–148. Clement, C.R.: 'A Center of Crop Genetic Diversity in Western Amazonia: A New Hypothesis of Indigenous Fruit-Crop Distribution,' *BioScience*, vol. 39, 1989, pp. 624-630. Denevan, W.M. and C. Padoch (eds.) *Swidden-Fallow Agroforestry in the Peruvian Amazon, Advances in Economic Botany*, vol. 5, 1988. Frickel, P.: 'Agricultura dos Índios Munduruků', *Boletim Do Museu Paraense Emilio Goeldi*, 1959 N.S. No. 4. Roosevelt, A. (ed.): *Amazonian Indians: From Prehistory to the Present – Anthropological Perspectives*, Tucson: University of Arizona Press 1994. Sponsel, L.E.: *Indigenous Peoples and the Future of Amazonia: An Ecological Anthropology of An Endangered World*, Tucson: University of Arizona Press 1995. Sponsel, L.E., Headland, T.N. and R.C. Bailey (eds.): *Tropical Deforestation: the Human Dimension*, New York: Columbia University Press 1996.
14 Anderson, A.B. *Alternatives to Deforestation Steps Toward Sustainable Use of the Amazon Rain Forest*, New York: Columbia University Press 1990. Balée, W.L.: 'The Culture of Amazonian Forests', in: D.A. Posey and W. Balée (eds.) *Resource Management in Amazonia: Indigenous and Folk Strategies, Advances in Economic Botany*, vol. 7 1989b, pp. 1–21. Irvine, D.: 'Succession Management and Resource Distribution in an Amazonian Rain Forest', in: Posey, D.A. and Balée, W. (eds.) *Resource Management in Amazonia: Indigenous and Folk Strategies. Advances in Economic Botany*, vol. 7, 1989, pp. 223–237. Redford, K.H. and C. Padoch (eds.): *Conservation of Neotropical Forests: Working From Traditional Resource Use*, Chicago: University of Chicago Press 1992.
15 Posey, D.A.: 'Contact Before Contact: Typology of Post-Colombian Interaction With Northern Kayapó of the Amazon Basin'. *Boletim de Museu Paraense Emilio Goeldi, Serie Antropologica*, vol. 3, 1987, pp. 135–154.
16 Posey, D.A.: 'Diachronic Ecotones and Anthropogenic Landscapes: Contesting the Consciousness of Conservation'. In: W. Balée (ed.): *Principles of Historical Ecology*, New York: Columbia University Press 1997.
17 Anderson, A.B. and D.A. Posey, in: D.A. Posey and W. Balée (eds.): *Resource Management in Amazonia: Indigenous and Folk Strategies. Advances in Economic Botany*, vol. 7, 1989, pp. 159–173.
18 Posey, D.A.: 'Indigenous Knowledge and Development: An Ideological Bridge to the Future?' *Ciência e Cultura*, vol. 35, 1983, pp. 877–894. Posey, D.A.: 'Indigenous Management of Tropical Forest Ecosystems: the Case of the Kayapó Indians of the Brazilian Amazon', *Agroforestry Systems*, vol. 3, 1985, pp. 139–158. Posey, D.A.: 'Environmental

and Social Implications of Pre- and Postcontact Situations on Brazilian Indians: the Kayapó and a New Amazonian Synthesis', in: A. Roosevelt (ed.): *Amazonian Indians: From Prehistory to the Present – Anthropological Perspectives,* Tucson: University of Arizona Press 1994, pp. 271-286.

19 These old fields are sometimes erroneously considered as 'abandoned' or 'fallows' by scientists, but this gives the false impression that they are unused and only waiting to become useful again for timber or agriculture.

20 Posey, D.A. and G. Dutfield: *Indigenous Peoples and Sustainability: Cases and Actions,* Utrecht: International Books 1997.

21 Chadwick, D.J. and J. Marsh (eds.): *Ethnobotany and the Search For New Drugs,* Chichester: John Wiley and Sons 1994. Joyce, C.: *Earthly Goods: Medicine-Hunting in the Rainforest,* New York, Toronto and London: Little, Brown and Co. 1994. Posey, D.A.: *Indigenous Peoples and Traditional Resource Rights: A Basis For Equitable Relationships?* Oxford: Green College For Environmental Policy and Understanding 1995. Reid, W.V., S.A. Laird, C.A. Meyer, R. Gamez, A. Sittenfeld, D.H. Janzen, M.A. Gollin and C. Juma (eds.): *Biodiversity Prospecting: Using Genetic Resources for Sustainable Development,* Washington, DC: World Resources Institute 1993.

22 Principe, P.P.: 'The Economic Significance of Plants and their Constituents as Drugs', in: H. Wagner, H. Hikino and N.R. Farnsworth (eds.): *Economic and Medicinal Plants Research, Volume 3,* London and San Diego: Academic Press 1989.

23 Posey, D.A.: 'Intellectual Property Rights and Just Compensation for Indigenous Knowledge', *Anthropology Today,* vol. 6, 1990, pp. 13-16.

24 Rural Advancement Foundation International: *Conserving Indigenous Knowledge: Integrating Two Systems of Innovation.* An Independent Study by the Rural Advancement Foundation International. Commissioned by the United Nations Development Programme. New York: UNDP 1994.

25 See Posey 1995. For results from the Workshop, contact: Green College Centre for Environmental Policy and Understanding, Green College, University of Oxford, Oxford OX2 6HG.

26 Research in which communities control research priorities based on their own criteria. These include self-demarcation, inventories of traditional resources, environmental/ social impact assessments, and resource management plans.

Protecting nature in Amazonia
Local knowledge as a counterpoint to globalization[1]

Maj-Lis Follér

Introduction

We are today facing initiatives on a global scale to support biodiversity, conservation and the protection of nature, and an interest in the local inhabitants, especially those living in tropical forests. International and national non-governmental organizations (NGOs), scientists and commercial companies have made new alliances with communities and native leaders and raised issues of biodiversity and protection. From grass-roots organizations and branches of social science ever greater attention is being directed toward democratic participation by local communities, intellectual property rights, and cultural and ethnic identity. Within the context of globalization the world is shrinking, and the dominant cultures, those of Europe and the United States, are penetrating the local world, e.g. the indigenous groups in the Amazon basin. In the process of globalization complex relationships between local activities and interaction across distances are developed.

This penetration of the Amazonian region will be discussed as follows. The indigenous people and the ecology of the region are placed in today's global context. The inhabitants and natural environment are discussed in terms of three prevailing images: *the image of superiority, the image of the ecological noble savage,* and *the image of the region as an empty space.* A historical ecology view of man's resource use is undertaken to show the long-standing man-environment interaction in the region, with a variety of forms of natural resource management. The globalization process with dominant economic activities is described to explicate its effect on local people and their subsistence pattern. This takes the form of an encounter between two sorts of knowledge – global scientific, and local knowledge.

The ecological upheaval

Global interest in ecological issues started in the mid-80's. The World Commission on Environment and Development – the Brundtland Report – was one of the first signals from the United Nations calling for action.[2] The concern became a focus of world-wide attention at the Earth Summit, the United Nations Ecology and Development meeting in Rio de Janeiro, in June 1992. This was one among several transnational mega-events sponsored by the United Nations, which have emerged as arenas for discussions and decisions on our common future. The implementation of Agenda 21 and the Convention on Biological Diversity by the nations of the globe was discussed during 1997 at the Commission's meeting in New York. How new power structures and hierarchies are created within today's world society has been analyzed with these meetings as a basis.[3] In the wake of the UN conferences, governmental agendas, international development and donor agencies, as well as national, international and transnational industries, have been involved in a green polishing policy. What these new green enterprises mean, the different incentives from the actors mentioned above, and the authenticity of the intention to make the planet more sustainable are difficult to ascertain. But there does seem to exist today a mainstream of political correctness regarding ecological concerns, and a demand that sustainable development, whatever this concept means, be part of the globalization process.

The Earth Summit took place in Rio de Janeiro in Brazil. This location might have been a contributing factor to the certain attention, mainly at the alternative meeting arranged by NGOs–the Global Forum, which was paid to the tropical forest of the Amazon region, and the right to land and survival of its native peoples.

The Amazonian Indian – our reflected image of the Other

Different images of the people inhabiting the Amazonian region have been constructed mainly by the Europeans since their first encounter with the region. I will, in this section, present some of these images and analyze the consequences they have had and still have for the development of the region.

The first one is *an image of superiority*. Starting from the sixteenth century we have written documents of the encounter between the conquerors, mostly Spaniards and Portuguese, and the indigenous populations. In *The conquest of America: The question of the other*, Tzvetan Todorov explores the encounter between two separate world views by interpreting how Christopher Columbus and the first conquerors were caught within their own social context (which, of course, the Indians also were). Todorov defines the Europeans' world-view as having three dimensions, expressed as the Natural (harmony between man and nature), the Human (culture, material wealth) and the Sacred. The natives' absence of clothes, which for Europeans symbolize culture, made them appear closer to the natural world, and they became as a part of the landscape, "... *somewhere between birds and trees.*"[4] As the Christian Columbus was, he also perceived them as pagans, without law and religion. He was only able to understand what he saw from his own cultural context in which clothes, wealth, and the expulsion of humans from Paradise stood for culture. From the beginning there was also a feeling of superiority, which engenders protectionist behaviour. Within this image falls a prevailing idea of the indigenous people as weak, poor, ignorant, infantile and helpless, waiting for someone to take care of their problems.

These images have been modified, but part of this perception persists even today. The fate of the region is more than ever before integrated into world society and the prevailing capitalistic system.

The second image can be seen as contradictory to the first one: *the image of the noble ecological savage.* The idea of the Amazonian people as living in harmony with nature is disseminated through newspapers, television programmes, spectacular visits by celebrities, etc. They have hereby become a symbol for saving the planet from ecological disaster. This image is grounded in a construction of indigenous peoples as being the true environmental conservationists. They have become the key symbols, and key participants, in the development of an ideology and organizational network that links Amazonian conflicts to international issues and social movements.[5] The problem is that this image is a Western projection of how we would like the indigenous peoples to be, and is not – just as with the superiority image – grounded in empirical studies. It hereby stereotypes the native people and idealizes them as non-destructive and practicing sustainable natural resource use.

This image has a strong symbolic value for the global component– for instance NGOs supporting rain forest peoples – of today's various forms of partnership between global environmental movements and local peoples. A charismatic Indian leader is selected and has to live up to the myth of how we want him to appear. As these organizations depend on voluntary contributions from supportive donors, the attractiveness of the idea of people living in harmony with nature is tempting.[6] The image of the noble ecological savage can also be exploited in commercial marketing, using the symbols of exoticism, naturalness and harmony with nature. This is a formula employed by firms like Body Shop, Shaman Pharmaceuticals and various ecotourism travel agencies.

The third image is that of *Amazonia as empty space.* This gives various actors the right to enter the region and act as if no one owned the land. One example of this is the political slogan "Land without people for people without land" which will be discussed later in this chapter. Other actors are defending large-scale projects where people are displaced because they see the region as a no man's land. The forest has also been seen as empty of human beings by scientists. Ecologists looking for study sites that would allow for examination of 'natural' processes uncontaminated by anthropogenic effects have hereby neglected the human history of the place.[7]

These three images can be taken as a simple way of outlining the views of different world actors with interests in the Amazon. The view of superiority is the all-pervading problem. Two investigators who have emphasized this in different ways are Arturo Escobar and Vandana Shiva. Escobar talks about regimes of representation, colonization of reality and dynamics of discourse in his strong effort to decontextualize the project of modernity and development. He wants to explore how certain representations became dominant and indelibly shaped the ways in which reality is imagined and acted upon at present. Shiva has the same approach in discussing the way science makes invisible the bearer of local knowledge. One of her examples is how the women in India for a long time have used the neem-tree for various purposes. Today, when the scientific community extracts substances from the tree in laboratories, it doesn't mention the discoveries made by Indian women through their application of local knowledge. Shiva makes this an instance of colonization of the mind.[8]

The activities of the various actors differ according to their incentives, which images they have of nature, and the degree of their motivation to act on behalf of the indigenous people. The economic activities taking place in the Amazon today are in the mainstream of what I mean by globalization, in the sense that people living in the tropical forest are affected by events and decisions taken far away from them. Their land, culture, identity, world view and knowledge are changing due to this globalization process. The invasion of the world system, with national, international and transnational actors into Amazonia, which is affecting the environment and the lives of its inhabitants, is, in my opinion, not only being accepted, but also met with a certain resistance. How strong this resistance is, e.g. in the form of using local knowledge for protecting nature, will be discussed as a *"counterpoint to globalization"*.

The three images can also provide an explanation of why the indigenous people, or 'primitives' as they have been called through history, were treated with brutality and inhumanity. Today an exploitation of them still exists, but a transition to exoticism and romanticism is also prevailing.

People and nature in the Amazon

The Amazon basin occupies roughly half of the South American continent, with a surface of six million square kilometres, and includes the Amazon river, the largest in the world. The man-environment interactions in this huge area, with many different ecosystems and vegetation types, will be elucidated to demonstrate the long time-span of human influence. Since prehistory, encounters between groups of diverse backgrounds, subsistence patterns, sizes and levels of dominance, have been taking place. Indigenous groups have lived there for millennia, and for hundreds of years the region has also been inhabited by a *mestizo* population.[9] Amazonia has been the site of migration, trade, war and exploitation, intensified with the arrival of the Europeans during the 16th century.

For the purposes of this article the dynamic of peoples' diverse subsistence patterns during history will be in focus. With the European conquest of the region, the transformations of man-environment relations were intensified. Fundamental shifts in the indigenous population's use of biological resources were one consequence. With a historical ecology approach, new light can be thrown upon why some groups changed from agriculture to hunting, violating the traditional evolutionary theory that postulates a development from hunting and gathering, to more advanced and intensified agriculture. This linear evolution has now been called into question among researchers. The Kayapó on the southern fringe of the Brazilian rain forest are one example of a group that has changed from agriculture to hunting.[10] Longitudinal studies from the Xavànte in Central Brazil also show that agricultural intensification has not been continuous over time. The Xavànte today devote less time to horticultural activities, and more to exploiting wild game than two decades ago.[11] For our understanding of today's Amazonian populations, their local knowledge and how they deal with the environment, it is crucial to explore the relationship between ecological degradation, subsistence changes and their integration into the market economy.

The meandering rivers on the slopes of the Andean mountains, such as e.g. the Ucayali river, change their courses over time. People living along the Ucayali and its tributaries, in utilizing the nutrient alluvial deposits, must move their houses and villages according to the natural changes of the stream.[12] Indigenous groups since the Amazonian Neolithic have altered the distribution of what are today useful resources. Balée states that what in the ecological literature has been described as virgin forest is an *"anthropogenic forest"*.[13] This process, which he names agricultural regression, shows the dialectic and dynamic changes of subsistence patterns. One crucial part of his research shows that plant resources of old fallows, more than of tall forests, constituted superior substitutes for domesticates lost long before.[14] The human footprints of the ancestors are today's subsistence basis.

Economic Activities in the Amazon

The dominant activities are consequences of the prevailing economic system. Nature has in this context received an instrumental value which is internalized in the idea of economic growth. Electricity and petroleum are highly appreciated commodities, and absolute preconditions for the existence of mega-cities and industries world-wide. In the same way cash-crop plantations and cattle farms are also in place to feed people living far away from the production. Activities like these need huge areas and they replace that which we *perceive* as virgin tropical forest with a variety of ecosystems, vegetation and water regimes. Anyhow, human activities have always had effects on the ecosystem. The question is, therefore, whether today's activities just differ in scale, or if we are now faced with an irreversible and thereby qualitatively different process?

Deforestation

Amazonia is today estimated to be the world's last great reserve of tropical timber. It has been protected by its sheer inaccessibility, as there have been no roads. In Brazil this changed first during 1965 with the opening of the Belém-Brasília highway in the eastern part. During the 1970s, the government of Médici realized that the Amazon could be used to release the pressure for agrarian reform in the Northeast. The image of the Amazon can be seen in the slogan of that time "Land without people for people without land".

Deforestation is a general concept for the diminishing of organic life in the forest, and in the tropical forest it is often seen as irreversible. Of significance for the future of this process are the ways different actors value the forest in terms of vital environmental services, economic resources (including medicines), and aesthetic value.[15]

What we have seen is that international and domestic timber companies flocked to the interior of the region with dramatic consequences for the forest. The deforestation rates have been very high. In the context of sustainability, deforestation is questionable, and the extinction of tree species is, according to biologists and ecologists, of a magnitude that is alarming for the biodiversity of the future.

Anyhow, commercial logging is just one cause of deforestation in Amazonia. By far the most important cause of deforestation in Brazil has been the conversion of forest to pasture-lands. These cattle farms are also a real source of profit for land speculation. About 85 percent of the deforestation in the Brazilian Amazon is caused by only 500 ranches.[16]

Mining projects like Carajas are other important causes of deforestation in Brazil. This enormous installation for mining iron smelts ore with the use of charcoal produced from the surrounding forest. Hydroelectric dams are also mega-constructions that change the landscape drastically wherever constructed, resulting in the displacement of people. They change the course of the rivers and inundate local peoples' living areas. In Brazil, the Balbina Reservoir, Jatapu Dam and the Cotinga Dam are examples of disputed projects.[17] They have blocked indigenous peoples' traditional fishing routes, altered the water quality in the rivers affecting the numbers of fish, and flooded inhabited areas, all with alarming effects on the ecosystems. Another problem is that the timber in the lakes formed behind the dams is not cut, but simply left to die and decompose. Most of the projects have been heavily criticized for neglectfulness in carrying out environmental impact assessments (EJA) or producing after-the-fact reports.[18]

Other questioned projects involve farming for cash crops where forest conversion is used both to establish massive monocrop plantations for agricultural produce, and to produce timber or paper pulp.

Consequences for the Indians

Again the project makers' image of the indigenous peoples and the environment can be seen. The Macuxi Indians living within the area of the Cotinga Dam project had their traditional communal living territory in the area which was supposed to be flooded. Not all the Indians with their land in the region reacted in the same way. One village under the influence of the Protestant church was in favour of the project. This group was described by the project leaders as *Indios do bem* (good Indians), while the Catholic influenced groups, who were in the majority and were against the project, were seen as "obstacles" to modernity.[19] In the project report the Indians are described as *"occupying the land"* and nothing is said about their right to the land, which is guaranteed in various international conventions.[20] This demonstrates the image of the Amazon as an empty space.

Modes of utilization

Two growing threats in Peru are petroleum prospecting and *coca* cultivation. As the cultivation and control of *coca*/cocaine is such a global issue, it is worth mentioning to demonstrate how outside economic interest in the region is a prevailing threat for biological diversity and the people. The degradation of ecosystems associated with the production of *coca* leaves, and their processing into cocaine paste, constitutes an enormous environmental issue.[21] The social effects on the native people are also very complex. In a short time perspective, they might gain more money cultivating *coca* than their traditional subsistence crops, and thereby be better able to support their families. In the long term, the magnitude of the threats to biodiversity is so great that displacement of communities is common, and, of course, the locals fall into a heavy dependence on the strong narcotics cartels and the illicit drug trade. In the Ucayali region in Peru, mostly *mestizo* villages have been involved in the cultivation, but there are also indigenous communities which have *coca* fields within their territories.

In Brazil the impacts of gold mining on human beings, especially the effect from mercury, as well as its penetration and accumulation in the ecosystems, have been highlighted.[22] The latest activity, with a different face, but still within the globalization process is ecotourism. To reach the wild and unspoiled countryside people from big cities all around the world leave their homes and transport themselves with aeroplanes many thousands of miles to reach the 'paradise'.

The efforts by environmental groups toward developing commercially viable and sustainable uses of the rain forest have led to the emergence of the new 'green entrepreneurs'. There seem to be benefits as well as pitfalls for the indigenous groups entering these partnerships. My intention is not to explore them more thoroughly here, but just to indicate what I am referring to. Some areas are: small-scale conservation projects where a local community plays an integrated part, pharmaceutical companies working in partnership with indigenous communities, other sustainable commercialization of renewable forest products, and ecotourism.[23]

Traditional ecological knowledge (TEK) as a counterpoint

The image of the noble ecological savage includes the idea that the local knowledge of the Amazonian population can be of great value for preserving the biodiversity of the forest in the future, and that the indigenous people are the equals of the conservationists. This is the position which the green movement advocates. Today's discourse seems to shift between the extremes of throwing the whole developmental paradigm away, thus just seeing the negative impacts of modernization, and, on the other hand, seeing the TEK paradigm as a magic solution for the future of the Amazon and a sustainable development. The two extremes have to be avoided, from a scientific point of view.[24]

TEK is often understood as being the opposite of scientific knowledge. It is characterized as a form of knowledge that has evolved throughout the centuries by learning from experience and living close to nature. It is region-specific knowledge related to daily activities like agriculture, natural resource management, human and animal health, and other such issues. But just as scientific knowledge does not comprise one single body, indigenous knowledge also has to be analyzed in a context, and not separately from the socio-cultural and ecological reality it is a part of. In science, as well as in TEK, there is good and bad knowledge (see Paul Richard's chapter). They are separate forms of knowledge, created in different social contexts and by people belonging to distinct power strata in world society.

Reflections on knowledge

From our image of superiority, Western scientists have tended to view peasants and smallholders in the rural areas of Africa, Asia and Latin America as having an inefficient and primitive technology, which is too labour-intensive and devastates the ecology leading to increased poverty. In the present transition to the image of the noble ecological savage, new approaches, especially concerning agrodiversity knowledge, are spreading in interdisciplinary settings to re-evaluate this negative view of smallholders and slash-and-burn peasants.[25] In these studies smallholders from various rural areas of the world are demonstrated to produce more per unit area than large farms in the same regions, and to do so with greater energy efficiency and less environmental degradation.[26] We also learn that slash-and-burn cultivation is not one system, but many hundreds or thousands of systems.[27] In "*Resource Management in Amazonia: Indigenous and Folk Strategies*", a broad variety of theoretical approaches dealing with local use, perception and manipulation of resources are presented, and case studies take the history of the people into consideration.[28]

The gap between the two images, that of superiority or what can be called the globalization paradigm on one hand, and the noble ecological savage or the local paradigm, on the other, is great. Both of them desire to achieve a sustainable future, one by economic growth, the other by strengthening the localities and stimulating alternative development with more self-sufficiency.

The question raised is, which kind of knowledge should be used to preserve a healthy environment and ensure the survival of the human species?

The sustainable development debate is about the state of the globe which we will hand over to future generations. The biodiversity debate among biologists is about which ecosystem should be given priority for protection. The rationale behind selecting the Amazon, for instance, as a prioritized region for conservation is heavily questioned by some scientists.[29] The application of the biodiversity debate to human beings, which suggests preserving endangered indigenous groups for genetic or linguistic purposes, is very controversial. There is, anyhow, a spreading debate on social diversity enclosing issues of cultural and spiritual values. Which values should be given priority for survival? And again, who can set these priorities?

The relationship between global environmentalism and indigenous groups

In more general terms, the vast backlash that the relationship between environmentalists and indigenous groups can lead to is due to the relationship being based on a symbolic representation. This defines authentic 'indianness' in ways that contradict the realities of many native people's lives, that is to say, it is a myth.[30] The risk is that the 'real' Indian is becoming invisible, and a model that moulds the Indians' interests to the needs and wishes of the organization is constructed.[31]

The market economy is penetrating the Amazon region with the dominant activities mentioned earlier such as logging, ranching, *coca* cultivation, etc., but even the green enterprises such as the Body Shop, Shaman Pharmaceutics and ecotourism demonstrate the need for many indigenous people to be integrated into the world society.[32] The assumption that Indians will always opt for long-term environmental conservation rather than short-term profits should be reconsidered, as well as whether the local communities really are interested in habitat conservation.[33] There are several examples in which native communities have assumed control over commercially valuable natural resources and chosen environmentally destructive options.[34]

The Kayapó Indians have had charismatic leaders such as Payakan, who was a media favourite until he fell from the pedestal for allegedly raping a white woman. They have appeared in advertisements for the Body Shop in feathers and body paints, and they have been exposed to the market economy and allowed timber companies to log virgin mahogany and other tropical hardwoods on their land. It is not my intention to paint the Kayapó Indians in dark colours, but rather to demonstrate how strong the market economy is, and the dazzling attraction of the society of materialism and consumption. The Kayapó's TEK, with intensive agriculture practices which increase the biodiversity in their plots and horticulture with buffer zones to increase yield, is not being questioned. But there are two strategies – continuing the traditional subsistence pattern, or being integrated into market economy – existing today within many indigenous communities, and neither of them is in-authentic.

In many articles TEK is abstracted from the history and reality where it was created, and seen in a functionalistic way. Also, expressions such as 'conserving' local knowledge gives an inactive impression, which should be avoided. The conclusions become a-historical and demonstrate a static view which I see as one of the traps in today's local knowledge discussion. Local knowledge is a complex cultural construction involving movements and events that are profoundly historical and relational.[35] A dynamic view can be found in Harold Brookfield's and Christine Padoch's article.[36] They point at the diversity of traditional resource-management practices and the many ways in which farmers use the natural diversity of the environment for production, including not only their choice of crops but also their management of land, water, and biota as a whole. They admit that this knowledge is under strong pressure, and is simplified due to economic circumstances, new crops and marketing opportunities, and growing population and demand. The important question to raise might be which conditions lead people to conserve their resources, and which favour destruction, or over exploitation of local resources.[37]

Conclusions

Biotic impoverishment is a reality today and biodiversity is a foundation of natural ecosystems. The trend of globalization is also seen as a natural law, but this is wrong. It is a politically structured ideology, produced in world society, within the prevailing economic growth theory. The effect of globalization on Amazonia, people and nature, has been discussed. Large-scale economic activities change the environment. The local inhabitants also destroy the rain forest with their extensive slash-and-burn agriculture and cultivation on steep slopes which causes erosion. Both of these are consequences of the globalization process–politics on an international or a national level.[38] Displacement of people is another consequence. People displaced from, e.g. a mountain ecosystem, bring with them their experiences, which might not be appropriate or easy to adapt to the tropical soils.[39]

Indigenous peoples before the globalization of Amazonia were dealing with natural resources in a manifold of ways. TEK might be challenging to include in today's economic and political processes. Context-specific indicators, and tests to measure those indicators have to be elaborated.[40] Subsequently, methods for integrating scientific and local knowledge in long-term sustainable solutions have to be implemented. The mono-cultures of the mind, as Vandana Shiva calls the ethnocentric view that local knowledge is biased and exploited by science, has to be revealed.[41] The myth of economic growth has to be questioned and the economic incentives to destroy the Amazon forest must be eliminated. In our academic reconceptualization of indigenous knowledge we have to avoid the complacent manner of speaking for them as a stereotype group who have asked us to speak for them. A reflection on the translatability of empirically created knowledge based on local events and objects into theoretical and general practice in new settings should be crucial to our research agendas in the future.

In reference to the title, Protecting Nature in Amazonia. Local knowledge as a counterpoint to globalization: First of all we have to reconsider what kind of world we are aiming for and thereby which nature we are protecting, and look for new alternatives for collaboration in an open-ended way. The pitfalls of constructing images and myths of reality also have to be considered. We must be aware of them. From a researcher's point of view the idea of scientific knowledge as being the only rational choice for a sustainable future must be questioned, just as other forms of knowledge should be scrutinized critically. A widening of science to cover other forms of knowledge together with less ethnocentrism might be a good start for the survival of Amazonia and its inhabitants.

Notes

1. This study was undertaken thanks to a fellowship as visiting professor from CNPq and Fiocruz in Rio de Janeiro September-December 1996. I also want to thank my colleagues Carlos E. A. Coimbra and Ricardo Santos at the Department of Endemic Diseases, Samuel Pessoa, National School of Public Health, Rio de Janeiro, who contributed with comments on an earlier version of this paper. This is also a part of a project called *Knowledge, Culture and Medicine* sponsored by the Swedish Council for Research in the Humanities and Social Sciences.
2. World Commission on Environment and Development: *Our Common Future*, Oxford: Oxford University Press 1987.
3. Little, P.E.: "Ritual, Power and Ethnography at the Rio Earth Summit", *Critique of Anthropology*, vol. 15 no. 3, 1995, p. 265.
4. Todorov, T.: *The Conquest of America, The Question of the Other*, New York: Harper Clophon Books 1982.
5. Conklin, B.A., and L.R. Graham: "The Shifting Middle Ground: Amazonian Indians and Eco-Politics", *American Anthropologist*, vol. 97 no. 4, 1995, pp. 695-710.
6. Ramos, A.: "The Hyperreal Indian", *Critique of Anthropology*, vol. 14 no. 2, 1994, pp. 153-171.
7. Redford, K.H.: "The Empty Forest", *Bioscience*, vol. 42 no. 6, 1992, p. 412-422.
8. Escobar, A.: *Encountering Development; The Making and the Unmaking of the Third World*, New Jersey: Princeton University Press 1995; Shiva, V.: *Monocultures of the Mind. Perspectives on Biodiversity and Biotechnology*, London: Zed Books Ltd 1993.
9. Posey, D.A.: "Protecting Indigenous Peoples' Rights to Biodiversity; People, Property,

and Bioprospecting", *Environment*, vol. 38 no. 8, 1996, p. 8. Posey has some definitions of who are the indigenous peoples. One of them is that they are considered indigenous if they are tribal peoples in countries where social, cultural, and economic conditions distinguish them from other sectors of the national community and where their status is regulated (wholly or partially) by their own customs, traditions, special laws, or regulations. In Peru the peasants are called *mestizo*, that will say a person with indigenous as well as European ancestors. In Brazil they also distinguish between *caboclos* and other, mostly peasant populations.

10 Balée, W.: "Historical Ecology of Amazonia", in L.E: Sponsel (ed.): *Indigenous Peoples & the Future of Amazonia*, Tucson: University of Arizona Press 1995, pp. 97-110.
11 Santos, V. R., N.M. Flowers, C.E.A. Coimbra Jr., S.A. Gugelmin: 'Tapirs, Tractors and Tapes: The Changing Economy of the Xavànte Indians of Central Brazil", *Human Ecology*, in press, 1997.
12 Follér, M-L.: *Environmental Changes and Human Health. A Study of the Shipibo-Conibo in Eastern Peru*. Humanekologiska Skrifter 8, Göteborg: Göteborg University 1990.
13 Balée, W., *op. cit.*
14 Balée, W.: *Footprints of the Forest: Ka'apor Ethnobotany – the Historical Ecology of Plant Utilization by an Amazonian People*, New York: Columbia University Press 1994.
15 Sponsel, L.E.: "Relationships among the World System, Indigenous Peoples, and Ecological Anthropology in the Endangered Amazon." In L.E. Sponsel 1995, reference 10, pp. 263-293.
16 Hecht, S.B. and A. Cockborn: *The Fate of the Forest. Developers, Destroyers and Defenders of the Amazon*, London: Verso 1989.
17 Fearnside, P.H. and R.I. Barbosa: "Political Benefits as Barriers to Assessment of Environmental Costs in Brazil's Amazonian Development Planning: The Example of the Jatapu Dam in Roraima", *Environmental Management*, vol. 20 no.5, 1996a, pp. 615-630; Fearnside, P.H. and R.I. Barbosa: "The Cotinga Dam as a Test of Brazil's System for Evaluating Proposed Developments in Amazonia", *Environmental Management*, vol. 20 no 5, 1996b, pp. 631-648; Fearnside, P.H.: "Brazil's Balbina Dam: Environment versus the Legacy of the Pharaohs in Amazonia", *Environmental Management*, vol. 113 no.4, 1989, pp. 401-423.
18 Fearnside and Barbosa 1996a, *op.cit.*
19 Fearnside and Barbosa 1996b, *op.cit.*, p. 644.
20 Fearnside and Barbosa 1996b, *op.cit.*, p. 646.
21 Young, K.R.: "Threats to Biological Diversity Caused by *Coca*/Cocaine Deforestation in Peru", *Environmental Conservation*, vol. 23 no 1, 1996, pp. 7-15.
22 Boischio, A.A. and D.S. Henshel: 'Methylmercury Exposure and Fish Lore among an Indigenous Population along the Madeira river, Amzon', in: M-L Follér and L.O. Hansson (eds.): *Human Ecology and Health. Adaptation to a Changing World*, Göteborg: Göteborg University, 1996, pp. 118-134.
23 Carr, T.A., H.L. Pedersen and S. Ramaswamy.: "Rain Forest Entrepreneurs: Cashing in on Conservation", *Environment*, vol.35 no.7, 1993, pp.13-15,33-38; King, S.R., T.J. Carlson and K. Moran: "Biological Diversity, Indigenous Knowledge, Drug Discovery and Intellectual Property Rights: Creating Reciprocity and Maintaining Relationships", *Journal of Ethnopharmacology*, no. 51, 1996, pp. 45-57; Fiallo, E.A. and S.K. Jacobson: "Local Communities and Protected Areas: Attitudes of Rural Residents Towards Con-

servation and Machalilla National Park, Ecuador", *Environmental Conservation*, vol. 22 no. 3, 1995, pp. 241-249; Stiles, D.: "Tribals and Trade: A Strategy for Cultural and Ecological Survival", *Ambio*, vol. 23 no. 2 1994, p.106; Tisdell, C.A.: "Issues in Conservation Including the Role of Local Communities", *Environmental Conservation*, vol. 22 no.3, 1995, pp. 216-228.
24 Escobar, *op.cit.*, p. 170.
25 Netting, R. McC: *Smallholders, Householders; Farm Families and the Ecology of Intensive, Sustainable Agriculture,* Stanford: Stanford University Press 1993; Brookfield, H.and C. Padoch: "Appreciating Agrodiversity, a Look at the Dynamism and Diversity of Indigenous Farming Practices", *Environment*, vol. 36 no.5, 1994, pp. 6-11, 37-45; Redford, K.H. and C. Padoch (eds.): *Conservation of Neotropical Forests: Working from Traditional Resource Use,* New York, Columbia University Press 1992.
26 Netting, *op.cit.*
27 Brookfield and Padoch, *op.cit.*
28 Posey, D.A. and W. Balée (eds.), *Resource Management in Amazonia: Indigenous and Folk Strategies*, Advances in Economic Botany, The New York Botanical Garden Bronx, New York: vol. 7, 1989.
29 Southgate, D., H.L. Clark: "Can Conservation Projects Save Biodiversity in South America?" *Ambio*, vol.22 no. 2-3, 1993, pp. 163-166.
30 Conklin and Graham, *op.cit.*
31 Ramos, *op.cit.*
32 Colchester, M.:' Indian Development in Amazonia: Risks and Strategies,' *The Ecologist*, vol. 19 no 6, 1989, pp. 249-254.
33 Southgate and Clark, *op.cit.*
34 Conklin and Graham, *op.cit.*
35 Escobar, *op.cit.*, p. 294.
36 Brookfield and Padoch, *op.cit.*
37 Redford and Padoch, *op.cit.*
38 Moran, E.F.: *Through Amazonian Eyes: The Human Ecology of Amazonian Populations*, Iowa City: University of Iowa Press 1993. Landless peasants from other parts of the country were offered land in the Brazilian Amazon.
39 Martinez, H.: *Migraciones Internas en el Peru,* Lima:IEP 1980. Governmental policy to open up the borders for large scale interventions have been mentioned earlier. In Peru, from 1940 to 1980, poverty drove at least a quarter of a million Andean farmers to settle in the Amazon.
40 Waltner-Toews, D.: "Thinking of Biology: Ecosystem Health – a Framework for Implementing Sustainability in Agriculture", *BioScience*, vol. 46 no. 9, 1996, pp. 686-689.
41 Shiva, *op.cit.*

Local peoples of the Western world – the introduction of local cultures in the Wadden Sea area

Ingeborg Svennevig

The Wadden Sea area in the North Western part of Europe fascinates both visitors, inhabitants and administrators. The diversity of land formations created by the forces of wind and water, the rich number of habitats and various species, the manifold sources affecting the area – all of this requires creativity and ingenuity if the natural processes in the area are to be preserved. In the preservation of nature both knowledge about the processes and about the human activities within them have to be combined.

The inhabitants' knowledge of the area is integrated in their practical activities, as expressed in the following description of the responsibilites of the 'count of the dike'[1] *"... the count of the dike must act on the spot, partly from his experiences with the weather and the water. A certain tone in the howling wind or a certain level of the waves compared to the time of day can show that 'now there is danger!' or that 'now it's changing – the danger has passed for now!'"*[2] The visitors on the other hand are introduced to the grandeurs of the area by demonstrating the consequences of their mere presence: *"If on a windy day on the beach, you stay put on the very same spot for a few minutes, as you then move your foot you will find that a small dune has formed itself. Here you can experience Nature at work."*[3] If we turn to the governmental reports, we are introduced to the area in a less experience-based manner: *"The Wadden Sea is a tidal area extending from Den Helder in the Netherlands along the coast and the islands of the German Bight to Esbjerg in Denmark. It is a network of tidal channels, sandbars, mudflats, salt marshes and islands spanning an area of about 900,000 ha and thereby constitutes the largest intertidal area in the world."*[4] These descriptions of the area can serve as indicators of the various perceptions that form the basis of nature preservation in the Wadden Sea area.

Nature preservation often implies difficult negotiations about how to rightfully distribute the resources between the interested parties: who is to decide, the farmer who wants to plough or the neighbour who enjoys the manifold flowers of a fallow field? There is a tendency to redirect this discussion away from the focus on resource rights and distribution, and reformulate it in terms of objective, environmental facts or indisputable, cultural norms. As an example of the culturalization of nature preservation, this paper analyzes selected self-descriptions of the inhabitants around the Ballum meadows in Denmark and on the island Texel in the Netherlands. My point being that in their attempts to secure their users' rights over the natural resources, the people in the areas introduce a special, local culture, different from the national, and thereby claim their rights to protecting their *cultural* peculiarity.

Indigenous peoples and the marsh inhabitants

By the references to a specific, local culture, different from the national, the inhabitants can be analyzed parallel to other indigenous peoples, and their struggle to gain self-determination within the overall system of nation states.

Though the history of territorial conquerings, followed by the subordination of the inhabitants is much older, indigenous peoples only became a general matter of concern after the Second World War. As the former colonized countries gradually became independent, the problems of the cultural diversity within the newly established states appeared. Until the beginning of the eighties, the indigenous peoples were characterized as being marginalized and oppressed, *because* they were culturally different from the ruling parties of their respective countries. An important factor in their marginalization was the lack of respect paid to their system of territorial rights by the state. The question of territorial attachment and rights has been and still is the key issue for indigenous peoples. The crucial point for them was that they would have the right to, but not necessarily get the opportunity to excercise, self-determination,[5] and this is still the basis for indigenous peoples' manifestations of themselves as distinctive from the complex of states – also when they are e.g. Western farmers, doctors, or lawyers characterized by living in a certain, demarcated area within the national territory.

Since the early eighties though, the political and economical aspects of being an indigenous people have receded into the background, and to gain public recognition, the campaigns for the oppressed peoples' rights have increasingly focused on their cultural uniqueness. The fight for the rights of the stateless peoples has now become international – or intercultural. There is a sympathetic understanding that it is the 'culture' of the indigenous that is the basis for their right to self-determination[6] – which means that if they cannot demonstrate a distinct culture, they lose their rights; and further, if their culture meets with no response interculturally, they will lose the international support.

An image of the indigenous peoples' culture that we often meet shows that they live in harmony with their natural environment. Both Indians in the Amazon rainforest, Inuits in the Canadian snow as well as bushmen in the Kalahari desert are confined within this image. The cultural image has gained independent life freed from the people it was supposed to illustrate. The result is not only that diverse cultures are stigmatized and perhaps homogenized, but also that the cultural manifestations are still harder to accept, except as peoples' strategic devices to gain support in the struggle for power.

To meet the quest for cultural authenticity, the continuity and originality of the cultural norms becomes crucial. When trying the authenticity of cultural traits, the inherent dynamic potential of a culture is judged negatively and concealed. Cultural norms seem to gain respectability the longer they have stayed the same. E.g. the hunters in the Ballum meadows try to create an understanding of the cultural norms guiding their hunting, but their claims of cultural continuity are met with suspicion, especially because of the technological inventions in hunting gear in the past generations.[7]

The situation of the marsh inhabitants is parallel to the remaining indigenous peoples of the world, in the sense that basically they feel that they are treated unjustly. In their opinion, the governments' cooperation with them as users' groups and local authorities does not render them sufficient influence, and therefore they appeal to their rights as distinct cultures. Though they do not dress in feathers or skins, they, like other indigenous peoples, will have their beliefs, trades, customs, family ties and ethical values scrutinized – and judged according to the observers' susceptibility. Though the marsh farmers are probably not as oppressed as other indigenous peoples, they are still trying to create a parallel case in their Danish and Dutch contexts: being accepted as distinct cultures with specific internal values and rights.

The marsh inhabitants and the national authorities

In the trilateral cooperation between Denmark, Germany and the Netherlands the Wadden Sea is preserved as an *ecological* entity, not as a political, national, cultural or economic one.[8] As in so many other cases of unique natural values, these do not confine themselves within humanly established borders, and the trilateral cooperation between the Wadden Sea nations is of paramount significance for the preservation of the Wadden Sea as, e.g., an eminent resting ground for migrating birds. Still, Denmark and the Netherlands use the international cooperation in different manners when negotiating with the inhabitants in the Wadden Sea area.

The Dutch government does not discuss international agreements with the local inhabitants; international agreements are integrated in the Dutch national policy. The Danish government, on the other hand, uses the international agreements much more explicitly, both to confirm the Danish self-perception of a pioneer country and as a reminder of our international obligations. Some Ballum inhabitants found it plain stupid to try to act as pioneers, others feared that the international level would disturb the fragile cooperation between government and citizens.

But it is still the national governments that are the sovereigns, and therefore held responsible by their citizens for their decisions, be they internationally obliged or inspired. The Danish national authorities were described by some of the marsh inhabitants as similar to the floods and storms in the low lying areas: extremely powerful, unavoidable, and redistributing the resources so that people would have difficult negotiations afterwards, when trying to agree on the proper distribution of the wreckage. Though this symbol is natural, the power relations between the parties were definitely not perceived in that way.

On Texel it was very difficult for the inhabitants to identify the national authorities. They described the government as having *many faces*, which is not so strange since the Wadden Sea is managed by at least four ministries, and an extensive system of institutional coordination.[9] The island's inhabitants described the preservation of nature as a conflict between different ideals within the same set of opposites, that were commonly agreed upon: ecology versus economy; nature versus agriculture. This cultural construction of conflicts appeared to me to be almost natural to the Dutch participants in nature protection.

The marsh inhabitants and the remaining national citizens

The marsh inhabitants agitated for their rights to be treated differently from the remaining citizens in their respective countries. Because they inhabited the beautiful Wadden Sea area, they had preserved the attachment to nature, which had been lost by others in the increasing urbanization. At present they felt that they were treated differently, but not in the right way.

The Wadden Sea area in the Netherlands is not as special and marginal as it is in both Germany and Denmark. Most Dutch have a sense of what it is like to build dikes, and thereby create and protect the land. Texelers preferred to present Texel as a *miniature of Holland*, offering everything that the mainland could offer – adding tranquillity and nature. Only after some afterthought (and my asking for it) would they remember that the mainland also had things to offer, e.g. agricultural subsidies, hospitals, etc. If anything, Texelers thought that they deserved positive discrimination.

In the Ballum area, which is both marginal and special compared to other places in Denmark, the Ballumers compared themselves to other Danes *and* to the rest of the Wadden Sea population, and concluded that they were living with the toughest restrictions. German and Dutch people were allowed to hunt – and apart from that their Wadden Sea areas were infested with heavy industries, so why restrict the few Danish small-scale users? And the rest of the Danish population had no idea what it was like to live with the risks of being flooded, or how it was to participate in conquering land and making it worth cultivating, so how could we justify making rules and regulations in this context? It was clear to the Ballum people that they were being negatively discriminated against.

In their respective, national contexts the Texelers and Ballumers seemed to have very little in common, their relations to their governments were oppositional in very different ways. But as marsh people they showed conspicuously similar traits when trying to express the uniqueness of their local culture.

Manifestations of a local, marsh culture

It is evident that all the people involved in the preservation of the Wadden Sea area feel the pride of this rich and dynamic wetland. The local inhabitants loudly appreciate their forefathers as the true protectors of the Wadden Sea area. They try to demonstrate that they have always taken good care of the Wadden Sea, and further, that preserving the values of the area has always had the their attention, otherwise, they argue, it would have been destroyed like the rest of the industrialized world has been. But they are suspect: are they not alarmingly close to transforming the natural processes in the Wadden Sea into economically sensible, cultivated, useful, human artefacts?

The inhabitants on Texel and in Ballum describe the foundation of the local culture as their dependency on the continual possibility to shape and re-shape the landscape in accordance with local values. They explain that the landscape functions as the model for human interactions: as a child one would learn to act correctly in the natural environment, and then one would unconsciously know how to act among other humans. In this manner nature would be protected and good citizens bred.

On Texel in 1997, the local party *Texels Belang*, with 6 out of 15 seats in the municipal council, presented the clearest description I have heard of the authentic local culture closely connected to the natural environment on the island: "*Century after century they [the original Texelers] have been experiencing the feeling of a certain freedom.*

Their independence in this matter is of great value and should on no account be ignored. After all it is their historical background and cultural history. (..) Over the years a huge amount of knowledge is gained by the local people regarding nature and landscape. (..) Knowledge and experience are transferable and therefore can be a sole interest for others. That way historical joint use will have an educational value."[10]

It is recognized that in the world of today, the 'historical joint use' (*historisch medegebruik*) is only an option for the chosen few. The authors appeal to the solidarity between humans and towards nature to accomplish the preservation of *these* exceptional humans. The manifestation concludes that *historisch medegebruik* is not some kind of favour that we will be doing the local culture, it is the wisest way to act under the circumstances. A type of human being which is closely connected to nature and possessing live, empirical knowledge, is about to be lost for ever, and when they are lost, we will all be *spectators*.[11]

The two representatives from the Texels Belang, who were both islanders for more than two generations, were proud to let me know, that their concept had been included, both in the municipal and in the provincial plans. Their hope was, that this would be an indication of a general change in the Dutch attitude towards local inhabitants in wonderful natural areas. Perhaps the authorities would finally have realized that if nature was to be preserved, they would have to redirect the attention from non-human species and their habitats, towards one kind of humans and their relation to the natural environment. The concept of *historisch medegebruik* challenges the ideal of nature untouched by humans, and can thus be introduced in the Dutch planning context as a substitution for the present concept of *nature development*; which means leaving people out, and waiting for nature to develop. Still, the *medegebruik* stays within the opposition between ecology and economy, nature and agriculture: *historisch medegebruik* is preservation of nature, it is not effective agriculture and it cannot pay economically.

The people along the Wadden Sea in Denmark also feel special, but they are not as explicit in their manifestations of a local culture. In the municipal councils, Bredebro and Skærbæk, it is still the national political parties that reign. In the Skærbæk council though, two members have joined forces, crossing political discrepancies, to fight what they perceive as a national attack on their local, cultural norms. In 1995, a hunting prohibition was a very controversial issue, and in a letter to the minister of Environment and Energy, one of them writes: "*Through the past generations, hunting and fishing have been an integrated part of our social and cultural way of life, and the children and young people of the area have quite naturally merged into this life style with their respect and use of the nature.*

Healthy and life-giving activities.

We strongly recommend that the hunting prohibition on the tidal flats owned by the state opposite the Juvre Dyb is cancelled, so that the way of life and activities of the local population are respected by the national authorities."[12]

Recently, the marsh inhabitants in the three countries have started coordinating their struggle against nature preservation. The 'local' aspect of their cultural manifestations forms the basis of their cooperation internationally.

The basis of the marsh peoples' culture

Up to this point, I have analyzed the local, cultural manifestations in the national and international context, where cultural manifestations have to be explicit and distinctive. Now I will turn to the internal aspects of this manifested culture, and analyze the significance of the landscape to the Ballum inhabitants.

When invited to cooperate on nature preservation, the 'locals' are considered to be people who feel a special attachment to the area that is to be protected. Nature protection is usually designated to the areas that we – outsiders – believe have preserved some of their natural characteristics – unmoulded by human interference. Against this stands the history of the inhabitants: the landscape in the Ballum meadows is considered to be Denmark's largest uninhabited area, but still the landscape is filled with symbols of name-given individuals. To the marsh people the meadows are shaped by Lorenzen, Christensen, Hansen and others. These individuals are depicted in the dike, the ditches, the windmills, etc. In the seemingly untouched area, the landscape can tell about numerous former conflicts and decisions, visible only to the trained eye. The landscape tells the culturally trained person about the ongoing negotiations concerning the beauty of the landscape; the cultural values of the marsh inhabitants; and the social positions of the honourable individuals that dared risk their family names on an innovation in the landscape.

It is only very special individuals that take the liberty to shape or reshape the landscape. When you feel convinced that your position allows you to make changes, you still have to act according to the values of the marsh people. E.g. one could imagine a person thinking that now was the right time to prohibit the cultivation of the meadows to protect some natural, dynamic processes there. This person, no matter how high a status he might have, would be confronting a value that has persisted in the area since time immemorial: man must conquer nature. It is still within living man's memory that the dike was built to protect humans, animals and crops. Older people also remember the first harvest in the meadows, it was richer than anyone could have dreamed of. Men in their forties remember their fathers cultivating the land with horses, etc. Together, all this results in a highly appreciated ability of humans to control the land for human benefit. Shaping the landscape in Ballum still has to be functional, directed towards securing the security and welfare of the inhabitants. And every choice a man makes will have significance, not only for his future relatives, but also for the family name as such, and thereby for the memory of his forefathers. In the local exchange of opinions about traces in the landscape, it can be relations between families that are at stake.

This locally based identity is confirmed and strengthened as to meet the government's invitation to cooperate on nature preservation. No culture, no matter how local it might seem, exists in isolation. The interactions with the surrounding world both inspires and forces cultural norms to be accentuated and perhaps changed. The following example illustrates my point: this year, 1997, the three Wadden Sea countries will probably agree upon a common management plan for the area. This has been a prolonged process between the national representatives, and they have reached historical decisions, one of them being that – except if there are human lives at stake – no more dikes will be built, and the existing dikes will remain the size they are now. Implicitly, the countries have now agreed to end their conquest of land, which is a major decision, at least in its Dutch context. At the same time, the Ballum Embellishment Society has unveiled a memorial pole, which evokes a living impression of the water levels during the past floods and storms in the area. With the pole's 5.5 metres, which is at the same time a token of the present constructor's emminent abilities since trees in their natural condition have never been able to stand this tall in the flat and windy areas around the Wadden Sea, the Society introduces the clearest possible symbol of the catastrophes the storms have brought to the area.

A local culture can be perceived as a mere reaction against national and international involvement in the management of an area. But this perception would neglect the internal aspects of culture specific norms and communication. They exist and persist because they render meaning and guidance in the worlds of the bearers. The implicit negotiations, that I have introduced here, present only one possible analysis of this internal aspect. As in the case of other indigenous peoples trying to manifest their cultural rights in other contexts, the marsh inhabitants have to find cultural traits that are susceptible to the surroundings in the present situation. Since this case of cooperation concerns the preservation of nature, this will also be the focal point for the people who want to express their cultural uniqueness. The continuation of a special, local culture is dependent on continuous shaping and reshaping of the area, in a dynamic relationship between human inhabitants and the landscape.

The landscape functions as a shared set of symbols of a local culture, and there is qualitative difference between being another interest organization and being a local culture – if they succeed in being recognized as such. As a cultural minority the marsh inhabitants will have the right to educate their children in accordance with their culture; and they will also have the right to fight for their cultural survival as a group of people connected by the landscape they inhabit.

Marsh inhabitants as nature protectors

Most people in Ballum – and in Denmark – smile at the marsh peoples' attempts to express the basis of a local culture explicitly. Obviously, the culture is dynamic and negotiable between the insiders, but it tends to gain its own imaginary life in relation to outsiders. Under the present situation, a local culture is defined in line with the remaining indigenous peoples of the world, and in this manner western 'indigenous' people try to enrol themselves as noble savages in the cross-cultural protection of nature. The inhabitants in the Wadden Sea area have chosen a culturally constructed counter strategy towards the measures meant to preserve nature. Can the local concept of nature correspond with the overall intention to protect natural values in a global perspective?

It seems that the technological development has created an immanent problem to nature protection with local participation in the Western world: people simply have too many and too far-reaching opportunities to alter the face values of the Wadden Sea area. Still, attitudes change, and farmers and their sons discuss the beauty of a fallow field, the function of windmills, the green colours of the sprouting crops, etc. The governments, on the other hand, try to stay on the narrow line between enforcing and stabilizing their laws, and still facilitating dynamic changes in the public opinion. The marsh inhabitants depend on the continued possibility to create a memory of themselves, and to involve themselves in the implicit discussions about the memory of name-given forefathers. The dynamic of this memory-creation might be possible to use as an integrated part of *nature* preservation.

The cultures of indigenous peoples would be less caricatured, if their explicit expressions were judged less as traditional, authentic traits, and rather as a precondition for being allowed the special rights of cultural minorities, and more as a dynamic possibility for peoples to act in a coherent world.

Notes

1 In the Danish Wadden Sea area, the majority of the dikes are maintained, with financial support from the government, by the people needing the protection. The count of the dike, *digegreven*, is the elected chairman of the group of people who pay the special fees to the maintenance of the dike, i.e. the people protected by it. The count is thus responsible for the regular operations concerning the dike, and for the emergencies in case of storms. Therefore, the *digegreve* is an highly influential person with a trusted assignment locally, considered to be a specialist in living in the low lying areas.
2 Fabricius, Nina & Dragsbo, Peter: *Ballum – et sogn ved Vadehavet*, Bredebro: Lokalhistorisk samling, 1996, p. 57 (my translation).
3 Fey, Toon: *Texel in het voetspoor van Jac. P. Thijsse*, Den Burg: Langeveld & De Rooy b.v., p.8 (my translation).
4 National Forest and Nature Agency, Denmark & The Common Wadden Sea Secretariat, Germany: *The Wadden Sea status and developments in an international perspective*, A Pijper Reprofessionals, p. 9.
5 Dahl, Jens: "Betragtninger over kategorien indfødte folk", *Antropologi*, no.32, 1996, pp. 25-33.
6 Veber, Hanne: "Indfødte folk og kultur. Moderne billeder og bevidste identiteter" ibid., pp. 57-68.
7 See e.g. Jepsen, Palle Uhd: *Strategi for en fremtidig jagt i vadehavsområdet*, National Forest and Nature Agency, January 27, 1997.
8 Common Wadden Sea Secretariat: *Forvaltningsplan for det trilaterale Vadehavssamarbejde, forhandlingsudgave*, February, 1997 & Enemark, Jens: "The Wadden Sea and the Dutch-German-Danish Wadden Sea cooperation in an agenda 21 perspective", paper presented at *Kystseminar*, University of Roskilde, 1996.
9 Zwiep, Karel van der: "The Netherlands", in Zwiep, K.v.d. & Backes, C. (eds): *Integrated System for Conservation of Marine Environments*, Baden-Baden: Nomos Verlagsgesellschaft, 1994, pp. 123-167.
10 Texels Belang, 1997, *Historical Joint Use*, information to the Province, unpublished.
11 Ibid.
12 Hunderup, Erik, letter to the Minister of Energy and Environment, 22.3.1995, unpublished, my translation.

Cross-Cultural
Conflicts and Cooperation

Modernity, nature and ethics[1]

Poul Pedersen

My article is provoked by an old joke, *"Why should we protect nature? Because there should be something to get back to!"* I do not mind that it is not the funniest joke in the world. What interests me is its assumption that we have left nature. I believe this is wrong. We have not left nature, we have only recently discovered it.

I am concerned with the global environmentalist discussion and, in particular, with the popular image of premodern or non-western peoples as gifted ecologists and dedicated conservationists. Typical examples of this are claims that ancient *"Hindu scriptures revealed a clear conception of the ecosystem,"*[2] or that various verses in the Qur'an express *"an elaboration of the concept of sustainable development,"*[3] or that *"the ethics taught by the [North American] Indian elders in sacred traditions is an ecological ethics."*[4] I will take a critical look at this greening of the 'noble savage' and argue that the assumed ecological correctness of premodern or non-western peoples is a projection of ideas which come out of a western knowledge tradition and that such ideas cannot be claimed to be authentic parts of a premodern or non-western tradition. I begin with a short history of modern western ideas of nature, and I focus particularly on time, space, quantification, meteorology, and systems of representation.

A short history of some modern Western ideas of nature

First, a few words about the word 'nature'. It has a bad reputation. To Leslie Stephen it was *"a word contrived in order to introduce as many equivocations as possible into all the theories, political, legal, artistic or literary, into which it enters."*[5] No wonder that a philosopher has recommended we avoid it and look for *"words of greater precision and stability of meaning."*[6] Here it would be too awkward to follow the philosopher's suggestion, but I will use the word 'nature' reluctantly and only in the "neutral" sense of the physical environment.

Modern knowledge about the environment is predominantly quantitative. It is characterized by universally defined systems of measurement and representation. People, things, and events are located within standardized and abstract global coordinates, grids, networks, and systems of time and space which are contexts for the interpretation of the world. This is not only an aspect of specialized, 'scientific' knowledge but also something which is part of most people's everyday life. In modernity, large-scale, global and universal categories have, in other words, turned into implicit frameworks for the individual and its relation to the environment. In this sense, the modern world is a globalized environment. This is the outcome of a long and uneven historical process: the transformation of the world by new technology, economy, and science which began in the Middle Ages and included the western global expansion and the industrial and communicative revolutions.[7]

In modern societies time and space are abstract and "external" categories.[8] In premodern societies they were integrated, "internal", aspects of localized modes of social life. Time was contained *in* localities. It was linked to the work on the land, to rituals, markets, and other local, social activities, and to the cycle of the seasons. Moreover, premodern localities were largely synchronically unrelated, life was organized around *"islands of time within seas of timelessness."*[9]

With the introduction of the mechanical clock from the late Middle Ages, time began to be separated from the local settings that had previously framed it. The clock measured an abstract and uniform dimension of time, which, in Mumford's words, *"by its essential nature dissociated time from human events"* – and *"human events from nature,"* as Landes has added.[10] Over the centuries time was increasingly "de-localized," and in the 19th and 20th centuries it turned into global or universal time with, first, the division of the world into time zones based on the prime meridian of Greenwich and, second, with the global standardization of calendars.[11]

The creation of universal or global time closely parallels the creation of universal or global space. In premodern societies space, like time, was related to localized social activities, which were in effect "place markers".[12] Long distance trade and migrations were not uncommon, but the locality was the dominant setting of social life. The centuries following the Renaissance saw a global or universal space being created above or beyond the localities as trade, travel, and communication expanded and spun a network of extensive social relationships between people with no shared common locality. The universal maps, which were produced in the wake of Western global expansion, and *"in which perspective played little part in the representation of geographical position and form, established space as 'independent' of any particular place or region."*[13] Over the centuries we can see a change in the relationship between time and space on the one hand and localities on the other. In premodernity, time and space are subordinated to localities, whereas in modernity, localities become subordinated to (global) time and space.

To the processes that turned time and space into abstract and global coordinates, should be added the close relationship between knowledge, production and quantification which played such an important part in the scientific revolution and which created totally new ideas about nature.[14] I can best illustrate the quantification of environmental knowledge by offering a few remarks on the history of meteorology and the introduction of instruments like the thermometer.

Nature quantified: the case of meteorology

The basic principle behind the thermometer had been known as far back as the third century B.C. when Philo of Byzantium described an apparatus which showed that air expanded or contracted with variations in temperature.[15] Philo and his contemporaries did not, however, see that this apparatus "really" was a thermometer, because they did not think of temperature (or, more generally, of the weather) as something to be measured or quantified.[16] They lived in the localized social setting of the premodern world, where knowledge about the weather was local knowledge, something founded on shared local experience. We get a sense of this concrete local notion of weather and climate in the work of, for example, the 10th century Arab scholar al-Muqqadasi, when he describes the climate at remote places by comparison with that of familiar places: "*This is a cold province, except Sijistan, Bust and Tabas-at-Tamr, where the climate is similar to that of the warm regions of Syria-Palestine. The climate of Bactria is Iraqi and that of Merv is Syrian. The winter of Khurasan is milder than that of Haytal. All of the province is dry but with variations from place to place. When the winter in this province is very cold, the summer will be very warm, except at Samarkand where the summer is pleasant, just like at Nishapur, where the winter, however, is milder than at Samarkand.*"[17]

In the centuries following the Renaissance, knowledge about the weather became increasingly separated from its previous local foundation and turned into global knowledge. The invention of meteorological instruments played an important part in that process, because they made it possible to quantify the weather, to break down its various aspects – temperature, atmospheric pressure, humidity, wind force etc. – into measurable units.

The thermometer was invented in the early 17th century.[18] The first thermometers were of the unsealed kind and not very useful because of their wide range of unconvertible scales of measurement and because they were subject to changes in atmospheric pressure. An important step forward came in the mid-17th century with the sealed thermometers based on the expansion of liquids that excluded the influence of atmospheric pressure. There was, however, a long way to go before heat and cold had become thoroughly quantified. Instruments had to be standardized and given interconvertible scales and the methods of observation should be defined and – not least – followed. This took time and involved the cooperation of observers, meteorologists, and scientific institutions on an increasingly international scale which resulted in the establishment during the 19th century of a dense global network of observation sites which even included ships crossing the oceans.[19]

By the end of the 19th century thermometers had spread to nearly all corners of the world making it possible for people to talk about and relate to a fundamental aspect of the environment, the weather, in terms of a standard and universally understood measure: the degree of temperature relative to some absolute 'zero' point. Helped by observations with thermometers and other meteorological instruments such as the hygrometer, the barometer, and the anemometer, the weather was broken down into quantified, comparable data. From the growth of controlled observation followed a new vision of weather and climate which reached far beyond the horizon of local time and space.[20] Described in the universal language of numbers and other symbols the weather now existed in a global space, where knowledge about it was increasingly separated from its previous pre-modern, local setting.[21]

Knowledge, representation, and ecological space

Measurements of natural phenomena were one important aspect of the quantification of nature. Another was modes of representation which made it possible to recognize global patterns of nature. One of the first to use graphic representation of complex information about the environment was Alexander von Humboldt.[22] His invention of the *isotherm* (in 1817) combined the global space of the universal map with the measured temperature of the global weather. Humboldt spent the years 1799-1804 in Central and South America. He brought along a large number of scientific instruments[23] and collected an enormous amount of data on geography, natural history, and economy, which he later published in 30 volumes. Humboldt was in line with the new trends in geography that aimed at a description of phenomena as they actually occur and coexist around us.[24] With an emphasis on the functional interrelation among all of the individual phenomena in nature, this went beyond the older Linnaean taxonomic idea of the *"system of nature."*[25] As a plant geographer Humboldt was not interested in arbitrarily isolated taxonomic entities, but in vegetation – in real natural wholes and their relationships to the total physical environment.[26]

From this perspective he pioneered the study of the geography of plants emphasizing how variations in vegetation depended on, for example, variations in altitude, temperature, and precipitation. By linking various kinds of environmental information he uncovered global patterns of nature and saw that it was made up of regional systems. The isotherm should be seen in this context.

The isotherm is a contour line which on a map links all places with the same average temperature. By bringing together geographical and meteorological information – and thus illuminating how geographical factors influence the prevailing temperature – the isotherm helped to initiate detailed studies of global and regional weather patterns. By drawing isotherms on a global map one could see that the distribution of continents and oceans had major effects upon climate – for example that the northern and southern hemispheres are not climatically symmetrical and that the interiors of continents have a greater variation of heat and cold than oceanic regions[27]

With the isotherm, variations in territorial shape and location were linked with climatic variation. Other modes of representation, for example maps of the distribution of species, could be related to climatic variation and made to illustrate other regional and global patterns of the environment. To take an example, one map with accompanying tables in Humboldt's *Essai sur la géographie de plantes* (1807) contains – in Humboldt's own words – information about *"the vegetation; the fauna; the geological connections; the agricultural cultivations; the temperature of the air; the limits of perpetual snow; the chemical constitution of the atmosphere [...] the horizontal refraction of sunlight, and the temperature of boiling water at different altitudes."*[28]

Humboldt shared with other Romantics a holistic vision of nature, but he differed from them in his dedicated empirical and quantifying studies. His efforts to represent nature as a system of quantified interrelationships had far-reaching consequences for, for example, meteorology. The traditional weather forecasts never went beyond weather rules.[29] The weather forecast as we know it today is the outcome of the transformation of time and space, of measurement and representation, and its predictive power is based on the knowledge of physical laws of causation and probability – on the quantification and modeling of nature.

Obviously, the environment is more than quantified interrelationships. It is, and has probably always been, a significant source of aesthetic experience and complex imagination. But what is new about modern environmental knowledge is its predominantly quantitative character which sets it apart from traditional knowledge and constitutes nature in an ecological space. This is the contrast between traditional and modern environmental knowledge that Glacken emphasized in his monumental survey of Western ideas of nature. He found that they formed a *"coherent body of thought"* in the *"time span from classical antiquity roughly to the end of the eighteenth century,"* which saw *"a closing, once and for all, of a period in the history of Western civilization."*[30]

A century ago T.H. Huxley argued that a *"new Nature begotten by science upon fact has pressed itself daily and hourly upon our attention, and has worked miracles that have modified the whole fashion of our lives."*[31] Huxley's *"new Nature"* is, in fact, the nature of modernity, and the knowledge of this *"new Nature"* has its exemplary expression in ecology which appeared as a recognizable science during the 1890s.[32] Studying the way living things interact with one another and with their environment, ecology observes, interprets, and represents thoroughly quantified natural phenomena. But this is only part of the picture, though a most significant one. Ecology is, besides being a quantitative science, also a concerned science. Or, perhaps one should say that it has become so. When the first systematic, ecological studies appeared a century ago, it had little to do with the environmentalist concerns we today associate with the word.[33] It was not until the late 1960s, with the growing public awareness of a threatening environmental crisis that it became common to think of ecology and nature protection as united in the mission to preserve the balance of nature. 'Ecology' became a global issue, and in the media the future was called *"the Age of Ecology."*[34] Today the word "nature" easily associates with "environmental crisis".

Radical environmentalists often associate modern environmental knowledge with western ideas of mastery of nature. Carolyn Merchant, for example, argues that *"for the past three hundred years, western mechanistic science and capitalism have viewed the earth as dead and inert, manipulable from outside, and exploitable for profits."*[35] One thing is that she makes a caricature of the sensible idea that science and economy influence attitudes toward nature, but much worse is that she totally ignores the significant role quantitative knowledge plays in, for example, the environmental ethics of the green movement where arguments for the moral responsibility toward nature are based on informed prediction about ongoing natural processes.[36]

To begin with I said that we have only recently discovered nature. I was, of course, referring to our ideas, our knowledge tradition, of nature, not to nature itself. The quantification of nature evolved slowly over the centuries and linked up with environmentalism only a few decades ago, turning nature into an ecological space. It is this recent merging of quantification and environmentalism I have called the discovery of nature, and the questions we ask about the environment after this cannot be separated from the knowledge tradition to which they belong.

Bear ceremonialism: environmental ethics?

I wish to emphasize that the simple opposition I have established between modern and premodern environmental knowledge is not meant to down-play the variation covered by each term. My intention is mainly to point out that they represent different knowledge traditions. With this in mind I will now discuss the greening of the noble savage. I take one celebrated example of assumed non-western environmental ethics, the bear ceremonialism as it is or was found among most of the northern hunting peoples of the Eurasian and North American continents. Here bears were highly respected animals and as hunting prey they were treated with special attention as was evident in the number of rites, ceremonies, and customs which were associated with the bear hunt. The first systematic study of bear ceremonialism was made by Hallowell in 1926.[37] He found that throughout the area it was common that bear hunters addressed the animal in kinship terms like 'grandfather' and requested it to allow itself to be killed. Hunters would also apologize for killing it, arguing that it was an act of necessity because they needed food or the skin for clothing. The killing was followed by various post mortem ceremonies and there were rules for the disposal of the remains. It has been said that the hunter-bear relationship is an exchange relation.[38] The hunters' respectful behavior toward the bear was reciprocated by the bear by allowing the hunters to kill it. In fact, it was not a direct relationship between hunter and bear. Hallowell refers to the belief that bears, like other natural species, have a chief, or spiritual master, whose orders they must obey. The respectful treatment of a dead bear is actually *"an offering or prayer to the chief of the bears to send more of his children to the Indians. If this were not done, the spirit of the bear would be offended and report the circumstances to the chief of the bears who would prevent the careless Indians from catching more."*[39]

In 1960 Hallowell took up the issue again. In his famous article on "Ojibwa ontology, behavior, and world view", he emphasized that the moral values which were implied in the Ojibwa veneration of bears *"document the consistency of the principle of mutual obligations which is inherent in all interaction with 'persons' [human as well as nonhuman] throughout the Ojibwa world."*[40] We should note here that Hallowell never referred to environmental ethics or ecology. His 1926 study was an exploration of the diffusion of culture traits, and the Ojibwa article from 1960 was an attempt to show how a culture functions as a whole. In the 1980s, however, bear ceremonialism was frequently interpreted as an expression of American Indian environmental ethics.[41] One who did so was the American environmental philosopher J. Baird Callicott.

Callicott seized on Hallowell's remark about the *mutual obligations*. But he created a context which was clearly different from Hallowell's. The implicit overall metaphysics of American Indian cultures, Callicott says, locates human beings *"in an environment in which reciprocal responsibilities and mutual obligations are taken for granted and assumed without question or reflection,"* and he continued, *"It is a world in which a person might feel at home, a relative to all that is, comfortable and secure – as one feels as a child in the midst of a large family."*[42] This is very different from what Hallowell tells us about the Ojibwa and their environment. They do not live a sheltered life. On the contrary, their existence is thoroughly marked by what could be called an 'ontological insecurity' which is rooted in precisely that close relationship they have to everything around them. Hallowell reports that his Ojibwa friends often cautioned him against judging by appearances, and he found in this advice an important clue to the understanding of their generalized attitudes toward the objects of their behavioral environment – and in particular toward people. The strong sense of the deceptiveness of appearances *"makes them cautious and suspicious in interpersonal relations of all kinds."*[43] In Hallowell's interpretation, this suspicion is a result of the deep-rooted belief that all 'persons', human or nonhuman, have transformational powers which they may use to gain control over other 'persons'. A 'person' may appear as someone or something else than it "really" is, and one cannot know whether that appearance is friendly or hostile. A smiling, amiable woman or a bear might be a sorcerer. The socialization process in Ojibwa culture which *"impresses the young with the concepts of transformation and of 'power' malign or benevolent, human or demonic,"*[44] aims at preparing the young for confronting the fundamental unpredictability in social relations.

All this has consequences for our understanding of *"the principle of mutual obligations which is inherent in all interaction with 'persons' throughout the Ojibwa world."* Certainly, mutual obligations were a comfort, but they were also a burden and a source of anxiety. Because social interaction was fundamentally unpredictable people could never be sure to have fulfilled all their obligations to other people or animals, and they had to be prepared for retaliation in terms of hostile acts like sorcery inflicted hunger or illness.

Callicott is wrong on the basic characteristics of Ojibwa culture when he ignores its deep-rooted anxiety. He also fails to explain why bear ceremonialism is an expression of environmental ethics. From all we know about traditional Ojibwa society it is difficult to say that bear ceremonialism worked as a restraint on bear killing. It seems much more in line with our knowledge if we think of bear ceremonialism as an anxiety reducing and risk diminishing technique which worked to the advantage of humans and against the animals because it made it possible to continue the killing of bears in spite of the severe risk it involved for the hunters and their society. In this perspective Callicott's interpretation of bear ceremonialism is a crude misrepresentation of the phenomenon.⁴⁵ When he praises the Native Americans for their ecological awareness, he is, in fact, inventing them in his own image. He is, while writing from a position where ideas of the environment are heavily influenced by ecological and environmentalist knowledge, crossing a huge cultural distance only to find the agenda of his own, modern knowledge tradition.

Callicott is here propagating what is often called "the myth of primitive ecological wisdom".⁴⁶ It belongs to a well established tradition of Western *Zivilisationskritik* which celebrates past or present non-Western peoples as morally superior beings. The myth most often serves a didactive purpose by assuming that we in the West can learn much about environmental ethics if we study non-western or traditional cultures. Callicott's example is hardly convincing in this respect, because his study of Native Americans has obviously not taught him anything he did not know before. In this sense the myth of primitive ecological wisdom is a threat to the heritage of *cultural diversity* because it represents the multitudes of traditional or non-western environmental ideas in the terms of a modern western knowledge tradition. Making us more ignorant of the variation of human lifeforms it results in a tragic loss of knowledge and sensibility – in some ways similar to the historical revisionism of totalitarian regimes.

Ethnic politics of ideas of nature

It is interesting to notice that appeals to traditional, religious values play a significant role in the global environmentalist discussion and that they often relate to concerns other than those about the environment. Since the '70s ideas of nature and religion have been closely associated with cultural identity, and in particular in the Third and Fourth worlds. Here political and intellectual elites have appropriated globally circulated ideas of ecology and environmentalism and merged them with the imagined, religiously sanctioned environmental ethics of their own imagined glorious, ecological past. The myth of primitive ecological wisdom has been taken over by people in the Third and Fourth worlds as a means of acquiring cultural significance in a globalized world. They demonstrate to themselves and to the world that their traditions, far from being obsolete and out of touch with modern reality, express a truth of urgent relevance for the future of the Earth.⁴⁷ However, there is a dark side to the celebration of primordial environmentalism. If ecological virtues are ethnic, or cultural, distinctions, environmentalist discourse may enter the dangerous field of ethnic conflicts, as is evident in the following example.

The Canadian-Indian professor of political science, O.P. Dwivedi, sees in the ancient Hindu scriptures "*a clear conception of the ecosystem,*" and he considers the caste system a *"progenitor of the concept of sustainable development."*[48] However, he faces a difficulty, because today India has severe environmental problems. Somehow the traditional environmental ethics of Hinduism lost is strength. How does he explain that? *"If, for some reason,"* he suggests, *"these noble values become displaced by other beliefs which are either thrust upon the society or transplanted from another culture through invasion, then the faith of the masses in the earlier cultural tradition is shaken."* He continues, *"That, it seems, is what happened in India during the 700 years of foreign cultural domination. The ancient educational system which taught respect for nature and reasons for its preservation was no longer available."* Dwivedi is, in other words, blaming the foreigners – the British and the Muslims – who invaded or conquered India. They are responsible for India's present spiritual and environmental degradation. We need not worry much about the British. They left India half a century ago. But we should remember that today there are more than 100 million Muslim citizens in India and that Hindu nationalists for a century have considered Muslims un-Indian elements. In this context Dwivedi's environmental concern is, with its Muslim-bashing, close to the xenophobic propaganda of the present day Hindu nationalists. The consequences of Dwivedi's analysis are frightening, and it is equally frightening to see his piece of Hindu chauvinism published in an otherwise decent book called *Ethics of Environment and Development*.[49] However, Dwivedi teaches us the lesson that *environmental ethics* is about more than *"noble values"*. It is also about *scholarly ethics*.

Conclusion

I have argued that the assumed ecological correctness of premodern or non-western peoples is a projection of ideas which come out of a western tradition of knowledge about nature and that such ideas cannot be claimed to be authentic parts of a premodern or non-western tradition as alleged by the myth of primitive ecological wisdom. Basically, it is an argument against a mode of explanation of human environmental behavior: the regression to pure, uncontextualized values. Non-western peoples are defined by their ideas of nature, their environmental ethics and values, and not by referring to what they do to their environment. This is contrasted with the notoriously bad environmental manners of the West. The result is a comparison of a real situation (the West) with an ideal, utopian fantasy (the Rest), and – consequently – a separation of people's values and ideas about the natural world from the way they are used in the interaction with the environment. The popular myth of primitive ecological wisdom is, also in its scholarly attire, promoted with the best of intentions, but unfortunately it obscures the complexity of people's real relationship to the environment – that they experience their environment in shifting and different contexts and rely on different and not necessarily consistent sets of ideas.[50]

We have only recently discovered nature, and the utopian myth of primitive ecological wisdom – with its confused longing for being at one with nature – is part of that experience. Perhaps we should listen to Woody Allen when he says, "*I am at two with nature.*"

Notes

1 I am grateful to Ingeborg Svennevig, Finn Arler, and Sally Laird for their kind assistance, and to Darrell Posey for one delightful comment, and to Don Lopez for providing me with certain bear facts. Part of this article is based on research carried out with Toni Huber of the University of Virginia (see Huber, T. and P. Pedersen: 'Meteorological knowledge and environmental ideas in traditional and modern societies: The case of Tibet,' *Journal of the Royal Anthropological Institute*, 2, 3, 1997). I am indebted to him for his outstanding scholarship and constant encouragement.

2 Dwivedi, O.P.: '*Satyagraha* for conservation: Awakening the spirit of Hinduism', in: Engel, J.R. and J.G. Engel (eds.): *Ethics of Environment and Development*, London: Bellhaven Press 1990, p. 205.

3 Izzi Deen (Samarrai), M.Y.: 'Islamic environmental ethics: Law and society,' in: Engel, J.R. and J.G. Engel: *op.cit.*

4 Hughes, J. D.: *American Indian Ecology*, El Paso, Tex.: Texas Western Press 1983, p. 81.

5 Cit. in Willey, B: *The Eighteenth Century Background. Studies in the Idea of Nature in the Thought of the Period*, London: Chatto and Windus 1946, p. 2. Cf. Williams, R.: 'Nature,' in his *Keywords. A Vocabulary of Culture and Society*, London: Fontana Press, 1988(1976), pp. 219-24, and Lovejoy, A.O.: 'Nature as aesthetic norm' in his *Essays in the History of Ideas*, Baltimore: Johns Hopkins Press 1948.

6 Hepburn, R.W.: 'Nature, philosophical ideas of,' in: Edwards, P. (ed.) *The Encyclopedia of Philosophy*, London: Collier-Macmillan, 1967, vol. 5, p. 457. The basic problem with 'nature' is, of course, not only that it is ambiguous but also that it is strongly normative.

7 Gellner, E.: 'The mightier pen? Edward Said and the double standards of inside-out colonialism,' *Times Literary Supplement*, February 19, 19??, pp. 3f.; Crosby, A.W.: *The Measure of Reality. Quantification and Western Society, 1250-1600*, Cambridge: Cambridge University Press 1997. Let me add here a few words about modernity and tradition. When I talk about modernity I think of various social forms which appeared in post-Renaissance Europe and in the following centuries became influential throughout the world. Industrialism, extended commodity production, nation-states, and an increasing application of science are important features of modernity. The movement from premodernity to modernity should not be understood in terms of evolutionary stages but as the uneven transformation of the world that has taken place over the last several centuries. In this perspective, there are no clear boundaries between traditional and modern societies.

8   Giddens, A.: *The Consequences of Modernity*, Stanford: Stanford University Press, 1990; Harvey, D.: *The Condition of Postmodernity*, Oxford: Blackwell.
9   Lash, S. and J. Urry: *Economies of Signs and Space*, London: Sage Publications, 1994, p. 227.
10 Mumford, L.: *Technics and Civilization*, London: Routledge and Kegan Paul 1934, p. 15; Landes, D.S.: *Revolution in Time: Clocks and the Making of the Modern World*, Cambridge, Mass.: The Bellknap Press of the Harvard University Press 1983, p. 16.
11 Dorn-Van Rossum, G.: *History of the Hour: Clocks and Modern Temporal Orders*, transl. by T. Dunlop, Chicago: University of Chicago Press, 1996; Kern, S.: *The Culture of Time and Space, 1880-1918*, London: Weidenfield and Nicolson 1983; Nguyen, D.T.: 'The spatialization of metric time,' *Time and Society*, 1, 1992, pp. 29-50; Zerubavel, E.: *Hidden Rhythms. Schedules and Calendars in Social Life*, Chicago: University of Chicago Press 1991. See also Sobel, D.: *Longitude*, London: Fourth Estate 1995.
12 Lash, S. and J. Urry: *Op.cit.*, p. 55.
13 Giddens, A.: *Op.cit.*, pp. 18f.
14 Richard Westfall has pointed to four aspects of the concept of nature which followed the scientific revolution of the 17th century. He writes that 'nature was quantified; it was mechanized; it was conceived to be other; it was secularized,' Westfall, R.S.: 'The scientific revolution of the seventeenth century: the construction of a new world view,' in Torrance, J. (ed.): *The Concept of Nature*, Oxford, Clarendon Press 1992, pp. 63-93. Here I am only concerned with the quantification, though the three other aspects are, of course, closely related to it. In this context it should, however, be noted that the quantification of nature was a slow process and that measurement and quantification did not take their fundamental role in the sciences until the 1830s and 1840s, cf. Hacking, I.: *Representing and Intervening. Introductory Topics in the Philosophy of Natural Science*, Cambridge: Cambridge University Press 1983, pp. 233ff; and Kuhn, T.S.: 'The function of measurement in modern physical science,' in his *The Essential Tension: Selected Studies in Scientific Tradition and Change*, Chicago: University of Chicago Press 1977, pp. 178-224.
15 Frisinger, H.: *The History of Meteorology: To 1800*, New York: Science History Publications 1977, p 47.
16 Or, in Alexandre Koyré's words, 'It is not so much the thermometer which is lacking, but rather the idea that heat is susceptible of exact measurement,' cit. in Cohen, H.: *The Scientific Revolution. A Historiographical Enquiry*, Chicago: Chicago University Press 1994, p. 87.
17 My transl. f rom the French in Miquel, A.: *La géographie humaine de monde musulman jusqu'au milieu du 11e siècle*, Paris: Mouton, 1980, vol. 3, p. 301.
18 Middleton, W.E.K.: *A History of the Thermometer and its Use in Meteorology*, Baltimore: The Johns Hopkins Press 1966. The inventor was probably Santorio. Frisinger gives the credit to Galileo and bases his claim on Middleton, *op.cit.* but appears to have misunderstood him, cf. Middleton, *op.cit.*, p. 14.
19 Feldman, T.S.: 'Late enlightenment meteorology,' in Frängsmyr, T, J.L. Heilbron, and R.E. Rider (eds.): *The Quantifying Spirit in the 18th Century*, Berkeley; University of California Press 1990; Frisinger: *Op.cit.*
20 Feldman, *op.cit.*, p. 177.
21 This did not, however, result in the disappearance of traditional knowledge about the

weather (weather rules like 'A red sky in the morning is the sailor's warning'). Many people probably to some extent still rely on traditional weather rules, but the authoritative voice on the weather is the meteorologist's. For an informative account of the 'culture' of TV weather programmes, see Ross, A.: 'The drought this time,' in his *Strange Weather*, London: Verso 1991, pp. 193-249. I am grateful to Peter Hansen for this reference.

22 Robinson, A.H. and H.M. Wallis: 'Humboldt's map of isothermal lines: A milestone in thematic cartography,' *Cartographic Journal*, 5, 1967, pp. 119-23.

23 Cannon, S.F.: *Science in Culture: The Early Victorian Period*, New York: Dawson and Science History Publications 1978.

24 Nicolson, M.: 'Alexander von Humboldt, Humboldtian science and the origins of the study of vegetation,' *History of Science*, 25, 1987, pp. 167-94; and 'Alexander von Humboldt and the geography of vegetation,' in Cunningham, A. and N. Jardine (eds.): *Romanticism and the Sciences*, Cambridge: Cambridge University Press 1990, pp. 169-85.

25 This change in scientific orientation was, of course, part of the picture Foucault has painted of the thoroughgoing transformation of the unconscious metastructures of western European thought, see Foucault, M.: *The Order of Things: An Archaeology of the Human Sciences*, New York 1970.

26 This is equally true of his interest in weather and climate. "*Die grosse Aufgabe, die sich Humboldt gestellt hatte, bestand in der Zusammenfassung aller meteorologischen Beobachtungen zum realen Klima,*" Schneider-Carius. K.: *Wetterkunde, Wetterforschung. Geschichte ihrer Probleme und Erkenntnisse in Dokumenten aus drei Jahrtausenden*, Freiburg-München: Verlag Karl Alber 1955, p. 151 (italics added).

27 Bowler, P.: *The Fontana History of the Environmental Sciences*, London: Fontana Press 1992, pp. 208f.

28 Cit. in Nicolson, M: 'Alexander von Humboldt and the geography of vegetation,' *op.cit.*, p. 181.

29 For examples of classical Greek and ancient Indian weather forecasting, see Theophrastus: *Enquiry into Plants and Minor Works on Odours and Weather Signs*, vol. 2 (transl. by Sir Arthur Hort), London: William Heinemann 1949 (*The Loeb Classical Library*); and Ramanathan, A.S.: *Weather Science in Ancient India*, Jaipur: Rajastan Patrika Limited 1993.

30 Glacken, C.J.: *Traces on the Rhodian Shore. Nature and Culture in Western Thought from Ancient Times to the End of the Eighteenth Century*, Berkeley: University of California Press 1990 [1967], pp. xii-xiii, 713.

31 Huxley, T.H.: *Method and Results*, London: Macmillan 1893, pp. 51f. It is not clear whether Huxley knew he was almost quoting John Dryden of 1668, "*Is it not evident, in these last hundred years when the study of Philosophy has been the business of all* Virtuosi *in* Christendome*) that almost a new Nature has been reveal'd to us? That more errours of the School have been detected, more useful experiments in Philosophy have been made, more Noble Secrets in Opticks, Medicine, Anatomy, Astronomy, discover'd, than in all those credulous and doting Ages from* Aristotle *to us? So true it is that nothing spreads more fast than Science, when rightly and generally cultivated*" John Dryden: *Of Dramatick Poesie: An Essay*, vol. 1, p. 12, in M. Summer's 1931 edition of Dryden: *The Dramatic Works*, cit. In Cohen: *Op.cit.*, p. 1.

Dryden saw what was coming, Huxley that it had come.

32 McIntosh, R.P.: *The Background of Ecology. Concept and Theory*, Cambridge: Cambridge University Press 1985, p. 27.
33 Bowler, P.J.: *The Fontana History of the Environmental Sciences*, London: Fontana Press 1992, pp. 361ff.
34 Worster, D.: *Nature's Economy. A History of Ecological Ideas*, Cambridge: Cambridge University Press 1991 (1977), pp. 339ff. On the globalization of ecology and environmentalism, see Pedersen, P.: 'Nature, religion, and cultural identity. The religious environmentalist paradigm,' in Kalland, A. and O. Bruun (eds.): *Asian Perceptions of Nature: A Critical Perspective*, London: Curzon Press 1995, pp. 258-276.
35 Merchant, C.: *Radical Ecology. The Search for a Livable World*, New York and London: Routledge 1992, pp. 41f., italics added ; see also her *Death of Nature: Women, Ecology, and the Scientific Revolution*, San Fransico: Harper and Row 1980.
36 Yearley, S.: 'Green ambivalence about science: legal-rational authority and the scientific legitimation of a social movement,' *British Journal of Sociology*, 1992, 43:3, pp. 511-32.
37 Hallowell, A. I.: 'Bear ceremonialism in the northern hemisphere,' *American Anthropologist*, vol. 28, no. 1, 1926.
38 See below.
39 *Ibid.* p. 69-70.
40 A.I. Hallowell: 'Ojibwa ontology, behavior, and world view,' in: Diamond, S. (ed.): *Culture in History. Essays in Honor of Paul Radin*, New York: Columbia University Press 1960, p. 47, italics added.
41 See, for example, Callicott, J.B.: 'Traditional American Indian and traditional Western European attitudes towards nature: An Overview,' in Elliot. R. and A. Gare (eds.): *Environmental Philosophy. A Collection of Readings*, London: The Open University Press, Milton Keynes 1983; Hughes, J.D.: *American Indian Ecology*, El Paso: Texas Western Press 1983; and Vecsey, C.: 'American Indian environmental religions,' in Vecsey, C. and R.W. Venables (eds.): *American Indian Environments. Ecological Issues in native American History*, Syracuse, N.Y.: Syracuse University Press 1980. An exception to these studies is provided by Jonathan Z. Smith's 'The bare facts of ritual' in his *Imagining Religion*, Chicago: The University of Chicago Press 1982, pp. 53-65. Smith says nothing about environmental ethics or ecology, but my argument owes much to his brilliant discussion of the ritual aspects of bear hunting.
42 Callicott: *Op.cit.*, p. 244.
43 Hollowell: *Op.cit.*, p. 40.
44 Dorson, R.M. *Bloodstoppers and Bearwalkers: Folk Traditions of the Upper Peninsula*, Cambridge, Mass.: Harvard University Press 1952, p. 30, cit. in Hollowell: *Op.cit.* p. 38.
45 It is tempting here to recall what Thomas Hobbes wrote in 1651 about mutual obligations between humans and animals, "*To make Covenant [agreement on mutual obligations] with bruit Beasts, is impossible; because not understanding our speech, they understand not, nor accept of any translation of Right; nor can translate any Right to another: and without mutual acceptation, there is no Covenant*" Tomas Hobbes: *Leviathan*, Harmondsworth: Penguin 1977 (1651) p. 197.
46 For critical discussions of this, see Ellen, R.: 'What Black Elk left unsaid: on the illusory images of Green primitivism,' *Anthropology Today*, 2, 6, 1985: 8-12, and Milton,

K.: *Environmentalism and Cultural Theory. Exploring the Role of Anthropology in Environmental Discourse,* London: Routledge 1996, ch. 4.
47 See note 35, Pedersen, P.
48 See note 2 for reference.
49 See note 2 for reference.
50 Croll, E. and D. Parkin: 'Cultural understandings of the environment,' in Croll, E. and D. Parkin (eds.): *Bush Base: Forrest Farm. Culture, Environmnet, and Development,* London: Routledge 1992, 11-36, p. 16; Ellen R.: 'Introduction,' in Ellen, R. and K. Fukai eds.: *Redefining Nature,* Oxford 1996, 1-36, p. 14; Norgaard, R.B.: 'Economics as mechanics and the demise of biological diversity,' *Economic Modelling,* 38, 1987, 107-21, p. 118.

Global partnership – a matter of friendship, reciprocity or mutual advantage?

Finn Arler

In the documents from the Earth Summit in Rio, reference to the "global partnership" is a recurring feature. In the last principle of the Rio-declaration it is proclaimed, that *"states and people shall cooperate in good faith and in a spirit of partnership"*. There is a need to *"strengthen friendly relations among states and people"*, the Convention on Biological Diversity says. And the Preamble to the Agenda 21 starts by declaring, that *"humanity stands at a defining moment in history"*, where sustainable development is only possible, if nations go together in a global partnership.

Literally speaking, there is nothing really new in these proclamations (although the term 'partnership' has not been used very often). The idea that humankind constitutes or should act as one common friendship of some kind, or even one large family, has a long history, probably with a religious origin (we are all children of God), and has been revived time and again in declarations and peace treaties. The modern global system is even based on that premise. Thus, the purpose stated in the first article of the UN Charter is to establish *"friendly relations between nations"*, and to *"harmonize their actions"*. And the 1948 Universal Declaration of Human Rights speaks (in the Preamble) about *"the human family"*, whose members should develop *"friendly relations"* and deal with common problems *"in a spirit of brotherhood"* (Article 1).

Another metaphor, however, which can also be found in the Preamble to the UN-Charter, is that of a *"global neighbourhood"*. To take part in a global neighbourhood is obviously less demanding than being part of a family, and it is not even as demanding as participation in a partnership. There is, of course, the biblical demand to love one's neighbour as oneself. In general, however, neighbours need not bother about each others' affairs as long as they do not interfere with each other, they have no kinship ties, and they need not have any more comprehensive common values and purposes.

This point is also reflected in the 1995 Report of the Commission on Global Governance, which uses the neighbourhood metaphor in its title: *"Neighbourhoods are defined by proximity. Geography rather than communal ties or shared values brings neighbours together. People might dislike their neighbours, they may fear or distrust them, and they may even try to ignore or avoid them."* In most neighbourhoods it is possible to move away, if the neighbours become too importunate and obtrusive. However, *"When the neighbourhood is the planet, moving to get away from bad neighbours is not an option."*[1] Many people prefer a neighbourhood, where the neighbours cooperate and make friendly ties, but neighbours need not get involved in each others' business. They need not share meanings nor sentiments. They may just be leaving each other alone. In a partnership this is too little, and even more so in families.

But how then should we understand global relations? Is there much more in it than just a proclaimed friendship? Can or should we hope for something more, and if so: what kind of friendship? Should we compare it with a family, a partnership, a neighbourhood, or is it more realistic to see it as a relationship, where fear, mutual distrust and hostility are so important features, that only the scantiest ties can be established, and only the weakest obligations be relevant? These are tough questions, and I shall not try to give definite answers. What I will do, however, is to put a little light on some of the premises. Firstly, I shall make a short sketch of how international relations have been interpreted in international law. And secondly, I shall present three different influential traditions of international ethics, in which some of the experiences of international relations have been summarized.

Global partnership in international law

a) The modern international system originated in the wake of the religious wars and revolts in Europe after the Reformation, and on the ruins of the sacred Roman Empire. In the Westphalian Peace Treaty (1648) ending the Thirty Years War, the basic idea of state sovereignty was explicitly stated, probably for the first time. It thereby officially confirmed an idea, which was formulated most clearly in Jean Bodin's *Six Books of the Commonwealth* from 1576, and which at the time of the treaty had been discussed for a long time in the international *republiques des lettres*.[2]

The purpose of the peace treaty was literally the same as that of the UN Charter, namely to establish *"Peace and Friendship"*, *"a perpetual, true, and sincere Amity"* in *"a good and Faithful Neighbourhood"* (Article I). The crucial move of the treaty was the establishment of sovereign territorial states (as heads of commonwealths).³ State sovereignty has two sides, which were clearly separated in the emerging new system. First, sovereignty on the outside was established as a right to defend the territorial borders of the commonwealth in order to secure *"full Jurisdiction within the enclosure of their Walls, and their Territory"* (Article LXVII), but also an external right to declare war, *ius in bello* (Article LXV). Secondly, internal sovereignty was based on a monopoly of force, *"the Right of direct Lordship and Sovereignty"* (Article LXXIII), strong enough to keep up not only with the Holy Roman Emperor, but also, and equally important, with all the different social powers, first of all the church(es) and the various factions of the nobility.

Although the treaty talks about universal and perpetual peace, the new international system was not really believed to make a final end to external war in cases of conflict. It was rather intended to keep warfare within reasonable limits, to avoid civic war between (religious) parties with incompatible conceptions of the good, and to help clearing up future conflicts by limiting the number of conflicting parties and the number of reasons for war. The states became *magna personae*, and the head of the sovereign state was legitimized as God's representative, and as such with unconditioned authority within state territory.[4] Each state, no matter how small, was formally accepted as equal to all others in rights and duties by analogy with the freedom and self-determination of private persons, pursuing their own self-defined goals.

Thus, international law became conceived as a law of co-ordination, based on neutral, formal and procedural norms, with a sharp distinction between the good and the right (territorial right, property, jurisdiction, tax-raising, access, acquisition, etc.). Each state (as head of a commonwealth) was sovereign in determining its own good; in international affairs, on the other hand, there was no place for the good, but only for neutral regulations of the right in the described sense. It should no longer be legitimate to make war because of religion or other kinds of beliefs. Thus, the establishment of a new interstate system could also be seen as a change of priority from unlimited moral conscience, rooted in religious (and therefore often uncompromising) conceptions of the good, to peaceful or at least more reasonable coexistence between representatives of a variety of beliefs.[5]

b) By the end of the 18th century the understanding of the commonwealth had changed as the third estate gained power, and the nation emerged as the new centre of concern. *"The principle of all sovereignty resides with the nation"*, says the French 1789 Declaration of the Rights of Man and of the Citizen (Article 3). *"We the people"*, the US Constitution begins. With the new understanding of the nation as the people's nation came a new understanding of social unity, based on secularized, culturally defined ideas of national identity (whether it be based on origin, language, history or pre-political *Volksgeist*), whereby replacing the earlier affiliations to particular ranks and classes.

All over Europe, in Northern America, and in various other parts of the world too, people became obsessed with more or less imaginative constructions of national identity, national tradition, national language, national history, national literature and painting, even national dress and recipes. But at the same time, and this should not be forgotten, they were involved in an actual establishment of common institutions and fora for cultural and political discussion and deliberation – and later on with national welfare systems, thereby creating more than just an imagined unity. The crucial change was the change of legitimation basis: sovereignty was grounded in the nation or the people, not in the will of God.[6] At the same time, however, this change lead to an internal tension between nationalism related to a pre-political conception of the people and its destiny, and democratic or republican conceptions of a common will crystallized through a democratic decision making process, based on private, political and (later on) social rights.[7]

The result of the changed conceptions was the 19th century system of nation states. What continued from the previous centuries was the understanding of the international system as a system of sovereign states, but now more clearly understood as a precarious balance of power (which had been disturbed by Napoleon's conquests). Therefore, the Chaumont Treaty from 1814, which established *la grande alliance* against Napoleonic France, had as its proclaimed purpose to re-establish the European balance between what was now called the Great Powers. Balancing the Great Powers was also the obvious goal at the peace negotiations during the Vienna Congress, even to the extend that smaller powers were treated as buffers and as bargaining counters. At least until the First World War there seems to have been a fairly strong consensus among decision makers, that the establishment of power balances – however precarious they may seem in times of crisis – was the only road to peace. For the same reason international law was only expected to have force among what was considered to be "civilized" nations, mainly within Europe: the right to intervene in "barbarous" states, with whom there was no genuine reciprocity, was left unrestrained, as the imperialist strivings throughout the century clearly show.

c) The creation of the United Nations in 1945 confirmed the established international system of sovereign nation states, the logic of the law of nations, which was spreading all over the world as former colonies became independent, and as the sovereign nation state tended to be the undisputed focal point of countries in almost all regions of the world. But at the same time it marked the ending of the earlier understanding of that system in at least two ways. Firstly, the UN Charter confirmed the renunciation of war,[8] which was stated for the first time (without much success, one may add) in the Kellogg-Briand Pact from 1928. It thereby finally withdraw the *ius in bello*, which had been a cornerstone in earlier conceptions of sovereignty, and revitalized Abbé St. Pierre and Immanuel Kant's old ideals of eternal peace.[9]

The UN Charter and the treaties which followed in its footsteps still reflected the competitive, potentially hostile setting, based on fear and mutual distrust. It therefore quite naturally focused on international peace and security. Self-determination and sovereign equality were keywords. But sovereignty now became more clearly limited by a set of basic rules in international law: nonintervention, renunciation of war, peaceful settlement of disputes, no threat and use of force, fulfillment in good faith of international obligations (Articles 1-2). The establishment of the UN, the Assembly and the Security Council and later on a series of other related organizations, meant that more force could be put behind the international treaties and agreements. The Charter explicitly mentions the use of military force as a possible solution, if a nation shows neglect of the law of nations (Article 42), but obviously, diplomacy and political and economic pulls and pressures have been much more important means.

Secondly, although national self-determination was the central norm (sovereign equality, territorial integrity, political independence), and international law therefore primarily a law of nations, sovereignty was now explicitly limited by a respect of human rights. This is stated already in the UN Charter, and repeated even more clearly in the 1948 UN Universal Declaration of Human Rights, which begins by proclaiming that *"recognition of the inherent dignity and of the equal and inalienable rights of all members of the human family is the foundation of freedom, justice and peace in the world"* (Article 2.3 and 2.4). The declaration even mentions the possibility of rebellion against tyranny and oppression, if human rights are not protected by the rule of law. But not (and here we find a clear asymmetry to the first limitation of sovereignty) the possibility of international military action against nations showing disrespect of human rights (UN Charter, Article 2.7).

d) In the first two decades following the Second World War, the common good was almost invariably expressed in the most neutral way: the central goals were peace, security, self-determination, and human rights to life, liberty and security. During the last two or three decades, however, we have witnessed how a still stronger priority is given to the common good conceived in a less neutral way. Thus, the notion of the good seems to re-enter the international system 3-400 years after it was excluded in the name of toleration and peace.[10] This probably reflects a changing international climate, in which hope and common deliberation slowly seem to be replacing fear and distrust as the basis of common regulations, and where problems of common concern of humankind can be put on the agenda. This change is probably clearer in environmental matters than anywhere else.[11]

Throughout this century a large number of bi- and multilateral treaties and declarations have been adopted, focusing on nature and the environment, especially the protection of wildlife. The first of them goes as far back as the end of last century. However, the 1972 Stockholm Conference on the Human Environment and the establishment of the UN Environmental Programme (UNEP) can be seen as the true beginning of international law on environmental issues – quite parallel, in fact, to the creation of environmental administration in almost all industrialized countries.

Once begun, the process has been extremely fast, and during the last 25 years more than one thousand bi- and multilateral treaties and agreements have been made on the regulation, conservation, preservation, and protection of natural and environmental goods and qualities. The general purposes of the efforts have been stated in a series of so-called soft law documents, among which the most important are the Stockholm Declaration (1972), the World Charter for Nature (1982), the Rio Convention (1992), and (less explicitly with an environmental focus) the Agenda 21 (1992). One could also add reports and strategy proposals from committees and organizations with a direct or indirect relation to the UN like the World Conservation Strategy (1980), Our Common Future (1987), Caring for the Earth (1991), and the previously mentioned report on Our Global Neighborhood (1995). These documents, reports and declarations are not legally binding in any strong sense, but they do spell out what the states believe the international relations ought to look like. In this sense they are closer to moral philosophy than to positive law.[12] Compared to the earlier rules, there seem to be four dimensions, in which an important change has occurred.

i) Pollution across borders has become a new issue. In the Stockholm Declaration it is stated clearly, that states have the responsibility to ensure that activities within their jurisdiction do not cause damage to the environment of other states (Principle 21). This is confirmed in Principle 2 of the Rio Declaration. The important thing to notice is, that damages caused by pollution are phenomena which cannot be formulated exclusively in neutral terms of the right in the previously described sense, i.e., in terms of property, territorial sovereignty and right of use and acquisition. Without an understanding of the good, damages are not possible to identify. This relation to more substantial conceptions of the good is even more obvious, when we look at the other three new dimensions.

ii) During the last three decades there have not only been made rules, but also more substantial goals for the international commons, as reflected, for instance, in the law of the sea, regulations of international territory (Antarctica), and regulations concerning the atmosphere.[13] For instance, an important part of the Law of the Sea is of course related to territorial rights, acquisitive and utilization rights to the resources of the sea, rights of innocent passage, etc. However, the provisions concerning conservation and preservation of species and other unique features related to the sea, or the prevention, reduction and control of pollution cannot be seen as neutral regulations, which can be stated independently of an understanding of the good. If there were no cross-cultural agreement on the need of protecting the goods in the sea, there would be no need to move beyond rules of territory and access.

iii) The concern for future generations has been expressed as a need for sustainable development. This is often described in fairly indeterminate terms at the international level, i.e. as far as possible in neutral terms of the right (e.g., future generations' equal right to self-determination, or equal access to environmental goods, or equal right to ecological security), but there is no way of avoiding a description of goods in more substantial terms, if we are to decide what more exactly should be left for our descending world citizens.[14] And this process of identifying goods is exactly what is taking place under seemingly neutral terms like 'natural and cultural heritage'.

iv) This is quite in line with a variety of more substantial provisions concerning the common good, including a series of recommendations which obviously limit or even interfere with national sovereignty. One could mention, for instance, the treaties and soft law documents concerning preservation of various kinds of biological diversity, endangered species, cultural and natural heritage, protection and improvement of environmental quality in threatened or vulnerable areas etc.[15] The provisions contained in the treaties are often quite distinctive. To give just one example: according to the Preamble to the Convention on Biodiversity, the contracting parties all agree to be *"conscious of the intrinsic value of biological diversity"*, as well as of its *"ecological, genetic, social, economic, scientific, educational, cultural, recreational and aesthetic value"*. Consequently, they all agree to be prepared to conserve the biological diversity *in-situ* within the territory under their jurisdiction, and to give various kinds of support to countries where lack of resources makes it difficult to comply with the agreement.

Preservation of biodiversity, protection of the climate system against dangerous anthropogenic interference, protection of the ozone layer, protection of vulnerable ecosystems at land and at sea, preservation of the natural and cultural heritage: these are all declared to be (in the language of the Earth Summit) common concerns of humankind. This may sound like one long story of succesful achievement. However, as everybody knows, the global treaties are all of a precarious nature, never stronger than the will of the governments in the signing nations. Some would even say that the governments are mainly paying lip service to the environmentalists, using vague phrases and avoiding specific obligations in order to keep everybody aboard.

There is no doubt that some governments are not doing much more than just signing declarations with no intention of compliance. What interests me, however, is not that some governments are not living up to what they have promised. I am only interested in the more general problem, how tight the global partnership ought to be, or how comprehensive it would be reasonable to think that international obligations should be. Most of us probably have some fairly mixed intuitions concerning this problem. In the following paragraphs I shall try to analyze some of these intuitions in connection with three typical approaches or traditions with different understandings of the problem. I shall argue, firstly, that all of them have at least one important point, which should not be forgotten. And secondly, that each tradition contains a brute as well as a refined version; when the three refined versions are put together, there may be some basis of consensus.

Three traditions of international ethics

There are, of course, a series of interpretations of trans- and international relations. I shall confine myself to a few secularized Western traditions of international ethics which have been very influential on the way the rest of the world interprets international relations, as more than anything else can be seen in the use of some key concepts. For instance, the sovereign nation state which until now has been the main unit in international law is a European invention, which has spread all over the world. This is also the case with the idea of human rights, as well as with concepts like power balance and *raison d'etat*. Global international law has to a very large extent been built on the European tradition of international law.

There is a dark side to this fact: European colonialization and imperialism. And it is easy to understand the distrust from former colonies and victims of European oppression toward what they often tend to see as just one more kind of imperialism, this time conceptual, dressed in only apparently universalistic terms. On the other hand, one of the positive reasons that Western traditions can be, and have been, useful in the interpretation of international relations, is that they all have one important experience in common: the presence of a variety of different and equally powerful cultural traditions, which have been forced to coexist one way or the other. (A fact which makes it clear that the notion 'Western' is in itself a precarious construct, covering a broad variety of traditions.)

This experience may not in all cases be followed up by a recognition that one's own way of living is not the only one, which can be seen as rational or even reasonable. But it does indicate that conflicts should be restricted if (self-)destructive wars are to be avoided. This is a lesson which has been learned the hard way from the disastrous events within European history, first of all the religious wars in the wake of the Reformation, and the two world wars in this century. With this in mind, let us now look at the three traditions in Western ethics.

a) Let me start at what may be called the particularist end of the spectrum. The rawest kind of particularism is undoubtedly the position of the Macchiavellian cynic, who sees the interstate system (and often the intrastate system, too) as a continuous battle field, where the strongest party tries to get advantages and goods, honour and wealth whenever possible through war and conquest.[16] The interstate system is understood as an instable and ruthless zero-sum game, where power alone determines the outcome, and where fear and distrust are the dominant feelings. If any political leader acted on moral grounds, he would soon lose everything, including his own life as well as whatever freedom and welfare he may have brought the people over which he ruled. Moral virtues turn out to be vices.

The Macchiavellian arguments still have influence in recent debate through the so-called "realist" tradition.[17] Modern realists may not be as straightforward as Macchiavelli in their misanthropic conception of human nature, and only very few of them would endorse Macchiavelli's recommendation always to strike first, but still, there are important similarities. First of all, international relations are seen as anarchic in an amoral sense, an actual or potential *bellum omnium in omnis* because of the lack of a common state authority. In international relations, power determines the rules, and each political unit, first of all the nation state, is therefore bound always to have security and the furthering of the narrowest national interests as its main, or rather as its only guiding principle.

Whether one feels comfortable about it or not, according to the realists, necessity rules out any kind of moral behaviour which goes beyond self-preservation and enhancement of the good life of the nation's own citizens (however this is understood). International affairs are not subject to questions of the common good, not even of justice, because beyond the national borders there are only balances of power and fragile contracts of mutual advantage.

Apart from the present preference for peace, however, there is one important difference between Macchiavelli and modern realism. Whereas for Macchiavelli the prince was the focal point of the theory, and the citizens were only cattle which could be mistreated whenever it was needed to keep the prince in power, modern realists are communitarian in the post 19th century sense: the focal point lies with the nation and the common, self-determined national good. Modern realists are accordingly more concerned about the heritage left for future generations, although only within the nation, and only as long as the nation exists with its specific set of symbols, values and ideas.

An even more refined kind of realism occurs, if the conceptions of the national interest include more than narrow-minded self-interest, i.e., if the common good is conceived in a way which includes higher level values like democracy and human rights, or preservation of various kinds of environmental goods for the descendants.[18] In fact, it seems difficult to exclude democratic values, if the national interest is to be more than a fancy tale. National interest is pure fiction if it is conceived without any relation to the common will of the people, and it would be quite impossible to identify this common will, if there were no democratic procedures, and if fear and insecurity were the dominant features. So realism may not only turn into communitarianism, it may even turn into a defence of moral values, which are not particularly related to a single community or nation, and which even seem to be at odds with the original assumption about sheer amorality beyond the national borders.

The important lesson to be learned from the (Macchiavellian and) realist positions is not the simple one, that every nation is bound to act selfishly (although, of course, there may be some truth in that), but rather that fear and lack of trust, combined with lack of knowledge and understanding are factors which should not be ignored, especially not in an international system without any common state authority. Whenever it is believed that other parties are acting on other premises than ourselves, trying to misuse whatever good will we may have, it certainly seems reasonable to be sceptical towards too ambitious ideas of global partnerships and human family ties. In that sense, there is definitely a need for at least some indications of reciprocity and a mutual recognition of sovereignty and self-determination.[19]

b) The realists' problem of identifying the national interest is only one branch of a general conflict within communitarianism. The crudest (and also the most dangerous) kind of communitarianism is the one which rather unreflectedly focuses on communal or ethnic or national *Volksgeist*. This is related to an idea according to which each nation (or ethnic group) has a unique, pre-political common spirit, expressed, for instance, in the national language, and more generally in the national culture, which is often believed to have grown organically out of a specific *Heimat*. In this kind of communitarianism the focus is on an authoritative national *ethos*, which leaves open only a few empty spaces for public dispute. It is static, backward-looking and authoritarian, often somewhat mystical and, in the worst cases, even expansionist.

There are few communitarians left, who would seriously defend such ideas in a critical discourse, and when this bad kind of communitarianism is still in use today, as we have seen in parts of Eastern Europe since the fall of the communist regimes, there often seems to be another agenda lying behind them, an agenda which is probably better understood in terms of power. Instead, modern communitarians typically underline the presence of a common continuous argument rather than an indisputable common *ethos*. In this case, the common ongoing history and the active political culture are underlined rather than a once-and-for-all determined common origin or descent. Confirmed and mutually recognized membership is considered more to be the issue than simple ethnicity. Consequently, the basis of legitimacy is seen in the democratically determined political expressions of the will of a people who conceives itself as such, rather than in pre-political myths of descent and destiny.

This does not mean that democratic values can make up the whole story of national identity. There is a variety of customs, practices, notions, institutions and features, which are unique to or come out in a special way in each nation. When we travel to a foreign country we immediately notice a great number of things which look different, and behaviour which would be considered odd at home. The local people refer to local heroes and villains, or to events we have never heard of, or which have no special importance to us. Because we are not part of their national community with all its implicit understandings, we feel a gap which can often be very difficult to bridge.

But still, a central part of the national identity could very well be related to a defence of more abstract, potentially universal values like democracy and human rights. And again, I find it very difficult to see how an authentic common will and an unrestrained social and political culture could be expressed, if these values were ignored. In the tension between nationalism related to a pre-political conception of the people and its destiny, and democratic or republican conceptions of a common will crystallized through democratic decision making processes, reflective communitarians would have to take side with the last party.

c) For those who are concerned about all the negative features of nationalism, this undoubtedly leads to the further question: why not jump right to cosmopolitanism, and leave the thicker conceptions of the good for individual decisions? After all, even within nations and communities there are many individual or sub-cultural lifestyles, some of which we may find almost as strange as those in foreign countries. Why then focus on more or less imaginary national or communal identities, in order to constitute partial relationships?

Just as in the case of realists and communitarians, cosmopolitan positions can be separated into the more and the less crude kinds. As before, let us start at the rawest end. To a true cosmopolitan, only individuals matter, not nations nor communities, maybe not even families. All individuals are people as such, and should be considered equally as such *in all matters*. Membership and identity should have no influence on our obligations: we should be impartial *in all affairs*. This could mean either that we should be impartial in all of our actions, so that we were always doing what would be best for mankind, never giving any kind of priority to, say, our own children. Or, that we should make rules and regulations, in which there were no place at all for membership and exclusive decision processes within limited social units, whether they be in families, communities or nations.

This kind of cosmopolitanism (which I believe to be close to what Brian Barry has called universal first-order impartiality), is implausible for a number of reasons. There are, for instance, the problem of compliance, the problem of identifying the common good *in all matters* despite the often huge cultural differences, problems of co-ordination, etc. etc.[20] Many of these difficulties seem to relate one way or another to one basic problem, which can be formulated in terms of thickness and thinness.[21] The closer we are, and the more we know about other people, the more understandings, goals and values we share with them, the thicker and more comprehensive our relationships can be. When we take part in families, we usually know and understand the other family members better than other people. Together we form common habits and practices, some of which can be difficult for outsiders to understand, and thereby also take on special obligations which cannot be identified in separation from the particular practices.

Similarly, we make close friendships with the special people we like and prefer to be with, feel loyalty towards, are concerned and passionate about, and whom we are willing to understand and defend a very long way.[22] We also form voluntary associations around some limited set of common interests and values, creating all kinds of more or less unique relationships with our co-members. And finally, we participate in communities and nations with a common sense of belonging to a collectively acting body of compatriots. A body, i.e., with a common history taking place within (although sometimes outside) a common homeland, and with a common public culture constituted by a continuous argument about standards of justice and goodness.[23]

In all these kinds of relationships we are partial in some sense. In fact, partiality is what families, friendships, associations, communities, and nations are all about. We find it quite natural to take on special obligations in each of these relationships, in which we know the rules and values, and often the specific persons involved. What is important to notice, is that without these special relationships we would undoubtedly be more shallow and less morally sensitive and considerate persons. We learn to take the place of others through close relationships, and only in this way can we develop our moral sensitivity. In this sense, appropriate partiality (which is indeed separable from phenomena like nepotism or chauvinism) must be considered a universally approvable feature, and it is one of the main reasons why local and national self-determination (within the described limits) is such an important value.

However, this also means that, in general, the further away in time and space (or to be more precise: culture) people live, and the less we know about them, share history, goals and understandings with them, the thinner our relationships as well as our obligations will be. In the end, there only seems to be some basic species characteristics, some fairly thin theories of the good, or some thinly described common denominators to appeal to. Unless, of course, an overlapping consensus is reached on thicker terms, as we have seen in relation to global environmental issues, but this already implies a higher degree of common understanding, and, as common policies are formed, a certain amount of common history.

A more modest, and therefore more plausible kind of cosmopolitanism would take such considerations into account. There seems to be different levels, on which cosmopolitanism or universal impartiality could be brought in. I have already mentioned, and rejected, cosmopolitanism at an everyday decision level. I do not find it any more convincing to say that all regulations should be immediately cosmopolitan, so that all borderlines and all strong commitments to particular communities should be rejected, and all regulations based on particular local understandings be denied.

The very recommendation that such commitments should not be suppressed does seem to contain a cosmopolitan or universal core, however. It is not only people belonging to our own community whose commitments should not be suppressed. Nobody's commitments should be suppressed. This second-order impartiality (to use Brian Barry's phrase) or second-order cosmopolitanism can be translated into political decisions concerning global partnership in various ways. One way is that of the law of nations as described above, especially when supplied with human rights guarantees.[24] One could also argue for responsibilities on humanitarian grounds, or argue for at least a minimum of transnational justice.[25] Following the considerations earlier in this article, I shall confine myself to still another interpretation, however, which seems to be of particular relevance in relation to the cross-cultural protection of nature and the environment.

As we have seen above, still more problems have to be decided on an inter- or transnational level. There is a variety of goods, which we cannot have if we are not acting as global partners at least on these issues. In this way our own history becomes intermingled with those of others. We have to make decisions in common, and we have to act on the basis of common agreements. This condition makes it all the more obvious, that the global decision making processes are not good enough to stand up to the big challenges they are faced with.

One of the fundamental problems is that the participants in the decision making processes are not all acting on the basis of the preconditions reflected, for instance, in the various human rights declarations. As argued earlier, these must be considered basic requirements for fair and mutually acceptable common history making (although the specific needs of each society may vary a lot, and therefore also the specific interpretation of the basic requirements). Only when they are fulfilled one way or the other, will it be possible to take common decisions among free and equal people at a transnational or global level, seeking agreement on reasonable terms. Reasonable solutions to common problems can hardly be expected to be within reach, if these requirements are not in place. In this sense, they are everybody's concern as world citizens.

There is one big advantage about taking the preconditions of reasonable common decision making as the focal point: the sharp distinction between (cosmopolitan) humanitarian or justice approaches on the one hand, and (realist) mutual advantage approaches on the other is weakened significantly. As participants in the common decision making process we should all be interested in fair conditions and procedures (in a broad sense). If this implies transfers of money and technology (as it undoubtedly does), it would not be the kind of one way street, which many people do not find justified in terms of distributive justice. Instead, it would be a mutual responsibility to enhance democratic procedures. The main responsibility would lie within the nation itself, of course, but the global partners would be committed to give a helping hand whenever it is badly needed. On the other hand, all obligations would lapse towards states with no intention of complying with this objective. Such a lack of compliance would be the ultimate sign of resignation from the global partnership.

Notes

1   *Our Global Neighbourhood*, The Report of The Commission on Global Governance (co-chaired by Ingvar Carlson and Shridath Ramphal), Oxford: Oxford University Press 1995, p. 44.
2   For a quick overview, see Murray Forsyth: "The Tradition of International Law" on the developments before this century, and Dorothy V. Jones: "The Declaratory Tradition in Modern International Law" on the changes in this century; both in: Terry Nardin & David R. Mapel: *Traditions of International Ethics*, Cambridge: Cambridge University Press 1992.
3   The treaty actually talks about reestablishment and confirmation of *"antient rights"*, whereby hiding the novelty of the system.

4   In order to avoid the church as the superior worldly power, the blessing of state sovereignty could not be mediated by the church. Thomas Hobbes gave the following explanation, which put the church as well as the people aside: *"The Office of the Sovereign, (be it a Monarch, or an Assembly,) consisteth in the end, for which he was trusted with the Sovereign Power, namely the procuration of the people; to which he is obliged by the Law of Nature, and to render an account thereof to God, the Author of that Law, and to none but him"* (*Leviathan*, ed. C.B. MacPherson, Harmondsworth: Penguin 1968, Part II, Chp. 30, p. 376). The church should only proclaim the Kingdom of Christ, which is not of this world, and therefore in no need of coercive power (p. 525).
5   The treatise thereby finally confirmed the principle *Cuius regio, eius religio* (the political ruler determines the official religion), which was stated for the first time in the 1555 Augsburg Religion Peace Treaty.
6   Miller, David: *On Nationality*, Oxford: Oxford University Press 1995, pp. 30f; Jürgen Habermas: "Der europäische Nationalstaat – Zu Vergangenheit und Zukunft von Souveränität und Staatsbürgerschaft", in: J. Habermas: *Die Einbeziehung des Anderen*, Frankfurt am Main: Suhrkamp 1996, pp. 135f.
7   Cf. Habermas, *op.cit.*, pp. 138ff.
8   Charter of the United Nations (1945), Article 2.3 and 2.4. The parallel articles in the charter of the League of Nations (1919) are much more vague; they only demand that the other parties of the League are consulted and used as mediators if a peace-threatening conflict emerges.
9   Including the immanent tension between *Völkerrecht* and *Weltbürgerrecht*, between the rights of nations and the rights of world citizens. Cf. Jürgen Habermas: "Kants Idee des ewigen Friedens – aus dem historishen Abstand von 200 Jahren", in: Habermas, *op.cit.*
10 It should be noted, though, that some basic conception of the good life is obviously present in the human rights declarations. The rights which are declared do after all reflect felt limitations of the good life. Therefore, we find among the declared rights a right to education, a right to form a family, a right to work and to rest and leisure, a right to adequate standards of living etc. A variety of human rights documents from all over the world can be found in: *Human rights in international law*, Strasbourg: Council of Europe Press, 1992.
11 For some good overviews of this remarkable change, see *Greening International Law*, ed. Philippe Sands, New York: The New Press 1994, and *Environmental change and international law: New challenges and dimensions*, ed. Edith Brown Weiss, Tokyo: United Nations University Press 1992.
12 Cf. Jones, *op.cit.*, pp. 42f.
13 Let me just mention a few of the most important conventions: The Oslo and London Ocean Dumping Conventions (1972), the UN Convention on the Law of the Sea (1982), Vienna Convention for the Protection of the Ozone Layer (1985), the Montreal Protocols (1987 and 1990), and the Framework Convention on Climate Change (1992).
14 Cf. Avner de-Shalit: *Why Posterity Matters*, London: Routledge 1995, and John O'Neill: *Ecology, Policy and Politics*, London: Routledge 1993.
15 Among the most important agreements are The Ramsar Convention on Wetlands (1971), the UNESCO Convention on Protection of the World Cultural and Natural Heritage (1972), the Washington Convention on International Trade in Endagered Species (1973), the Bonn Convention on the Conservation of Migratory Species (1979),

the Basel Convention on Transboundary Movements of Hazardous Waste (1989), the Forest Convention (1992), and the Biodiversity Convention (1992).
16 Macchiavelli, Niccolõ: *The Prince*, Chicago: University of Chicago Press 1985. It should be noted, that the inclusion of points from his other discourses would bring forward a less cynical picture of Macchiavelli.
17 For an overview of realist positions, see Steven Forde: "Classical Realism" and Jack Donnelly: "Twentieth-Century Realism", in: Nardin & Mapel, *op.cit.*
18 The North Atlantic Treaty is a good example of this mixture of idealism and realism. The objectives are to *"safeguard the freedom, common heritage and civilization"* of the peoples of the contracting parties, *"founded on the principles of democracy, individual liberty and the rule of law"*. The means are military force.
19 Cf. John Rawls: *Political Liberalism*, New York: Columbia University Press 1993, pp. 16ff.
20 For a variety of arguments against this first-order cosmopolitanism, see for instance Marilyn Friedman: *What Are Friends For?*, New York: Cornell University Press 1993; Miller, *op.cit.*; and Brian Barry: *Justice as Impartiality*, Oxford: Oxford University Press 1995.
21 Cf. Michael Walzer: *Thick and Thin. Moral Argument at Home and Abroad*, Notre Dame/London: University of Notre Dame Press 1994. Walzer has borrowed the terms from Clifford Geertz: *The Interpretation of Cultures*, New York: Basic Books 1973. The thick/thin metaphor can also be found in John Rawls' distinction between thick and thin theories of the good, cf. John Rawls: *A Theory of Justice*, Oxford: Oxford University Press 1972, pp. 395ff.
22 The partiality of friendship has been a common theme in socalled feminist ethics, cf. Friedman *op.cit.*, and Virginia Held (ed.): *Justice and Care. Essential Readings in Feminist Ethics*, Boulder, Colorado: Westview Press 1995.
23 Cf. the discussion of national identity in Miller, *op.cit.*, chp. 2.
24 The law of nations position can be found, for instance, in John Rawls' *A Theory of Justice, op.cit.* He has now supplied this with an insistence on human rights guaranties in "The Law of Peoples", in: Stephen Shute & Susan Hurley (ed.): *On Human Rights, The Oxford Amnesty Lectures 1993*, New York: Basic Books 1993.
25 For a variety of such arguments, see Robin Attfield & Barry Wilkins (ed.): *International Justice and the Third World*, London: Routledge 1992, and Onora O'Neill: *Faces of Hunger*, London: Allen & Unwin 1986.

Religions and conservation. A survey

Tim Jensen

Contemporary religion, in a wide variety of expressions, plays an equally wide variety of roles in local and cross-cultural environmentalism and in discourses on environmentalism.

Studies, relevant to the academic study of the history of religions, mostly have focused on the eclectic and syncretistic environmentalist interpretations and applictions of so-called primitive religions and Eastern religions. Discussions are mostly centered on the question of the scientific truth of these (modern or New Age) interpretations and the possible political and ethical consequences of the application of what has been termed the "myth of primitive ecological wisdom".

Consequently, the studies imply renewed anthropological fieldwork among living indigenous peoples and revitalized reflections on past theories of the place and concept of what today is called 'nature' and 'environment' in the respective cosmologies and practices. Past and present modes of cultural, religious and academic 'primitivism' and the ways contemporary 'primitive cultures' (or groups among indigenous people) adopt and use the environmentalist interpretations are also being scrutinized. Sometimes, like in the case of Kay Milton or anthropologists working with development projects among indigenous peoples, the studies combine "pure" scientific interest in verification or falsification of the myth with an interest in clearing the ground for an improved understanding and handling of environmentalist issues.[1]

Few scholars, however, have addressed the issue of this survey, that is, the organized participation in environmentalism and conservation by groups within the so-called world religions,[2] especially the projects of those groups within the institutionalized world religions that cooperate with a non-religious environmentalist organization, the WWF. The survey is based upon analyses of the written statements, rituals, festivals and conservation projects as well as upon analyses of the classical material, mainly that which is referred to by the participating groups themselves.[3]

The religio-historical context and frame of reference, for the participation of the religions and the analysis of this, is the post – World War II process of globalization, the recognition of some crises as global, its impact on the religions and the reactions from the religions. Characteristic reactions have been: 1) A thematizing of the religious traditions, especially of the ethical traditions or themes relevant to the global issues and crises. 2) An increased effort to reenter national and international fora for discussions on cross-cultural and global issues, like poverty, over-population, genetic engineering, the environmental crisis. 3) The setting up of more fora for inter-religious, cross-cultural and ecumenical communication and activities. A study of the greening of the religions, consequently, contributes to broader attudies in these contemporary developments, and it must include questions of how these developments can be seen as indications of a "re-politicization" of religion and de-secularization of politics.[4]

The theoretical and methodological framework of the survey is founded on an open-ended, operational understanding and definition of religion as a historical and cultural variable.[5] Religions, in this perspective, are human, social and cultural constructs and interpretive systems, transmitting and processing, or, to quote A.W. Geertz, "producing and reproducing, unity, meaning, and meaningful relationships even in the face of inconsistencies". The vitality, validity and vindication of the religious interpretive systems depend on their capacities to adjust to historical change and challenges while simultaneously referring to their past traditions and to a postulated transfalsifiable, transhistorical and transnatural reality.

According to this anti-essentialist understanding, religions are dynamic systems of interpretations and religion cannot be separated from the human beings, the religious communities and individuals, who use it and apply it to the present situation. The history of religions is, from this point of view, the history of the religious interpretations of religion, and the task of the scholar of religion is the analyses, descriptions and explanatory interpretations of the processes of transformation and syncretism, of tradition and renewal and of the many instances of the 'reinvention of tradition'.[6] The process of "the greening of the world religions" must be placed within this general frame of reference.

Critics evaluating the current greening of the religions regularly point out the inconsistencies between the classical not-so-green and the modern much-more-green looks and outlook of the religions, noticing that religions 'originally' did not entail a cosmology or soteriology correlated to an environmentally benign concept of nature, and that, even if they did, a benign impact on the environment would be hard identify and hard to prove.[7] The perspectives and theories of the scholar of religion may be useful in this context. Stressing that there is no eternal essence to religion and to the single religions, that the use and continuous reinterpretations of religious traditions cannot be separated from the 'religions themselves' and that change and transformation is the rule, not the exception, may help critics from falling victims to some (almost religious) ideas about some eternal core and essence of the religions, and it may further the rational, scientific and humanistic understanding of religion as well as the interdisciplinary understanding of cross-cultural environmental issues.

The alliance of religions and conservation

> "One day in 1953 two men stood on the summit of Mt. Everest, Sir Edmund Hillary, a Western scientist and Sherpa Tenzing, a Himalayan Buddhist. Separated as they were by culture and beliefs, they had together scaled the highest mountain in the world and had, for the first time in history, reached its summit. What they did speaks volumes for the real differences between them and their cultures. Edmund Hillary stuck a Union Jack, the flag of Great Britain, in the snow and claimed to have 'conquered' Mt. Everest. Sherpa Tenzing sank to his knees and asked forgiveness of the gods of the mountain for having disturbed them."[8]

In an almost mythical, and certainly striking and paradigmatic way, this story, told by Martin Palmer, the mastermind behind the WWF inspired religious environmentalism, in an introductory essay on the Alliance of Religions and Conservation (ARC), opens up vistas for the understanding of what he calls "the practice of conservation by religions". In a less poetic form based upon explicit statements and analysis of relevant material, the philosophy and reasoning behind the WWF project of forging an alliance of religions and an alliance of religions and conservation can be summarized as follows:
– The environmental crisis is, though not exclusively, due to a mental and moral crisis, a worldview, predominantly but not exclusively, Western and Christian, characterized by a materialistic, dualistic, antropocentric and utilitarian relationship to nature.
– The environmental organizations fighting the crisis are to a large degree victims of the same worldview and the same economic and technological thinking that contributed to the crisis.
– Alternative worldviews and ethics, alternative ways of seeing man and nature, must be introduced and combined with the technological and economical resources.
– The world religions, no matter if they had nothing to say about the crisis before asked, no matter how they might have contributed to the crisis (encom-

passing environmentally malign ways of seeing and using nature or not acting according to environmentally benign ways of seeing nature), constitute enormous, environmentally benign, human and spiritual potentials.

The first fruits of this reasoning were the Assisi-event. In 1986 WWF chose to celebrate its 25th Anniversary inviting representatives from the 5 major world religions – Buddhism, Christianity, Hinduism, Islam and Judaism – to join secular conservationists in a pilgrimage to the birthplace of St. Francis, who (in 1979) had been officially elevated to the rank of patron saint for ecologists. In a way characteristic to many projects following this, the representatives from the religious, scientific and environmentalist communities turned an environmental meeting into a pilgrimage and vice versa. Having dramatized the destruction of many a first people and third world environment by way of formal apologies to a Maori-warrior obstructing the entrance to the basilica, the participants ceremonially proceeded into the basilica. After elaborate, specially designed ceremonies, they finally produced the Assisi-declarations, stating their past neglect of the environmental issues, their present goodwill and the environmentally benign potentials of their religious traditions. 'The New Alliance' or the 'Network on Religion and Conservation' was established.

The cross-cultural aspects and the 'unity in diversity' was touched upon in the introductory speech given by the Minister General of the Franciscans: "We are convinced of the inestimable value of our respective traditions and of what they can offer to reestablish ecological harmony; but at the same time, we are humble enough to learn from each other. The very richness of our diversity lends strength to our shared concern and responsibility for our Planet Earth."[9]

Over the next few years four other religions joined (the Bahàis in 1987, the Sikhs in 1989, the Jains in 1991 and the Taoists in 1995), each making a written statement and beginning action programmes. The Assisi-event was followed by inter-religious events in Canterbury, Washington, DC and Copenhagen. In 1995 thousands of religiously based environmental projects were running worldwide, some assisted by WWF, some springing directly from local sources. In 1995 the United Nations commented that the programme of working with major religions had reached "untold millions" with the conservation message, and in 1995 two more meetings took place, the one in Japan and the other, "The World Summit of Religion and Conservation", in England. Representatives from the nine religions came up with a joint commitment and revised editions of their former declarations, and another international structure, the ARC, was created, new programmes designed and old projects evaluated.[10] According to the latest newsletter from the ARC, the work has now (1997) inspired the World Bank to initiate consultations with representatives from the religions and the consultants of the WWF.[11]

Religiously based projects

There is no reason to mention the hundreds of projects (e.g. reforestation-projects, sustainable clerical administration, educational schemes, environmental prizes, organic farming, etc.) differing from secular ones only because initiated and run by members of a religious community. It is, I think, more interesting to look at examples of projects with a more evident and exclusively religious flavour:

1. Reinterpretations and declarations

In order to come up with an integrated and compelling environmental ethics linked to their stipulated central religious aims and soteriology, the religious groups have reworked central aspects of their various cosmologies, theologies, anthropologies and ethics, sometimes in the forms of elaborate publications, sometimes in the form of shorter declarations. Mostly they issue each their declaration, but sometimes they also succeed to demonstrate the 'unity in diversity' by way of issuing common declarations (in Copenhagen in 1990 and in Japan in 1995).

2. Inter-religious meetings.

Christians from various cultures and denominations have for many years gathered under the wings of the ecumenical organisation, the WCC, joining the JPIC-programme (Justice, Peace and the Integrity of Creation). The other religions, though on a smaller scale, have had their internal ecumenical meetings or tried to make various groups contribute to joint-venture projects. More significant however, are the international, inter-religious conferences on religion and conservation, where representatives from the various world religions and from various parts of the world gather. Though joint-venture inter-religious conservation projects do exist, the conferences primarily serve to signal that the religions recognize their responsibilities and the global character of the crisis and the need to cooperate across doctrinal, religious or cultural boundaries.

3. Celebrations of traditional festivals and rituals expressing the environmental outlook and responsibility

Special inter-religious events, like inter-religious services in famous churches, have been celebrated, and even special, cross-religious rituals have been designed to honour the unity in diversity. The diversity, however, and the ritual counterpart to the reinterpretation of the texts, have been displayed and celebrated in several festivals and rituals within each religion. Festivals and rituals traditionally linked to the seasons and to symbols referring to the natural as well as the

supernatural have been used to (re-)introduce awareness and responsibility for the environment. Sometimes the celebrations are linked to practical projects of conservation. A few examples:
- Traditional Christian Harvest-festivals have been redesigned and reinvented in order to make participants remember the 'gift of God', the obligation to share the fruits of the earth with the poor, and the neglected responsibility towards the creation of God.
- The Christian Mass (Eucharist, Communion) has been celebrated and re-interpreted in order to stress that the doctrine of the Incarnation and final Resurrection implies more than the human being, namely the sanctification of the whole creation and the resurrection of the cosmos. Bread, wine and Christ are infused with cosmic sacredness and additional significance. The celebration of the Mass signifies the death and resurrection with Christ to the benefit and salvation of the entire creation.
- The Jewish Sabbath has been celebrated to underscore the cosmic and recreative implications, and several other Jewish festivals (Shavuot, Sukkot, Tu Bishvat) have been used to strengthen the bonds between man, nature and God and linked to active reforestation programmes.
- Hindus celebrating the god Krishna during the pilgrimage to his sacred forests of Vrindavan have combined this with a tree-planting project. The participants entering the sacred precinct are handed trees to plant and given information on the deplorable status of the birthplace of the god.

4. Restoration and conservation of Holy Land
Several of the conservation programmes linked to land considered sacred, reveal that the religious concepts and practices sometimes have contributed to the crisis and that this fact is acknowledged. This goes for the above mentioned Vrindavan-project where millions of pilgrims for centuries have celebrated their god at the expense of the forest, and it goes for the sacred river Ganges which – due to the very ideas of its absolute purity and to polluting ritual practices – has become a veritable sewer. Today, however, some Hindus try to persuade the holy men and merchants dealing with the cremations to combine the two opposed concepts of purity and to minimize the devastating use of wood for the funeral pyres. Religious ideas and practices are also partly responsible for the present need to restore the Taoist and Buddhist sacred mountains of China, though general pollution and mass-tourism have played a part too.

A recent off-shoot of this kind of religious environmentalism is the large-scale British "Sacred Land-Project". The aim is to rediscover, reopen, reinvigorate and replant old pilgrimage routes and other sacred spaces. A project linking up with the many projects of turning Christian churchyards and Muslim cemeteries into ecologically sound spaces.[12]

Common denominators, and devices in their own right, are the deliberate use of the traditional veneration for the sacred, of respect for the religious authorities and of the communicative potentials embedded in religious language and symbols. A Buddhist monk may succeed in preventing villagers from cutting down the forest if he tells them about the bad karmic implications of the act. He may, however, get a better result erecting a statue of the Buddha at the spot, nailing parts of his yellow robe to the tree-trunks or insisting on the presence of tree-spirits. These devices are all species of a common cross-cultural and cross-religious 'Divine Command Ethics' and a globalization of the concept of nature and the environment. Yet, the exact means of communication and understanding are bound to the tradition in question and accommodated to the various targets and target-groups.

Common key-concepts

Turning from the projects to the motivating worldviews, values and ethics, analyses of the sources reveal that certain notions are highlighted. In general, however, the declarations refer to virtually all the central ideas in the cosmological and ethical systems.[13]

A slight idea of the notions, religious ideas and traditions, highlighted by the reinterpretive work, may be gained from the following (incomplete) list based on an arrangement of the religions in 'families':

I. Buddhism and Hinduism

a) *Mysticism*. The perception of the relationship between man and nature is deeply influenced by the mystic traditions referring to experiences and ideas of totality, oneness and unity.

b) *Pantheism*. The notion that the 'ultimate reality', 'truth' or the divine pervades everything finds its classical Hindu expression in the idea that the eternal innermost being of man, *atman*, and the innermost universal being, *brahman*, is one and the same. In Buddhism a similar notion is expressed in concepts of the all-pervading 'emptiness' or 'Buddha-nature'.

c) *Ahimsa*. Since man and nature are, in the last analysis and final liberating insight, made of the same 'soul', 'buddha-nature' or 'emptiness', the ideal of absolute and total, active non-violence (ahimsa) must be central. Due to ideas of karma and samsara and the central (especially Buddhist) notion of the interrelatedness of all things, everybody and everything must be treated with care and compassion.

d) *Karma-samsara* and *reincarnation*. The idea of eternal cycles of rebirths, on the universal as well as the individual level, and the idea that the thoughts and actions of man determine the mode and way of future lives, plays – in spite of all differences concerning the notion of soul (atman and anatman) – a central role in connection with environmental ethics.

e) *The world as a cosmos.* The world is conceived of as well-organized totality, a cosmos, where each and every part has its place. It is part of man's place and religious duty, and incorporated in the scheme for attaining final liberation, to help uphold the cosmic order by playing his role.

f) *Ignorance is a sin.* Ignorance is a sin leading to wrong and detrimental conceptions and actions. It is often stated that the environmental crisis is due to man's ignorance and false conceptions of the true relationship between the (falsely perceived) 'ego-self', the 'buddha-nature' and the environment.

Christianity, Islam and Judaism

a) *God as Creator.* The notion of the world as the creation of a transcendent and eternal god. The religions all deal with the 'problem' of the created world as different from, but given by, God, and in Christianity the idea of The Holy Trinity and the Incarnation adds a special flavour to the discussions. In all three religions traditions of the mystics convey notions bridging the gap between the transcendent God, man and the created world, notions that at times get close to pantheism or to pan-en-theism.

b) *The world as cosmos and man's place in the cosmos.* The concept of the created world as a cosmos is the basis for the discourse on the 'Integrity of Creation'. Man *is* something special and differs from the rest of the creation and the 'fruits of the earth' are there to be used by man. However, contrary to previous, dominant interpretations, especially of the Biblical texts, the stress today is on stewardship instead of dominance.

d) *Destroying the world is a sin.* The destruction of the God-given nature, consequently, is a religious and moral fault, a sin. This notion links up with the notion of the original fall as a consequence of man's arrogance.

The similarities within the families of the religions have to do with historical developments and past processes of syncretism. The similarities across the families, however, must be explained otherwise and cannot be explained by cross-fertilization at the modern inter-religious meetings of the religions. For the present purpose it suffices to mention a few of the most relevant explanations:
- In all the religions man relates to nature and the environment vis-á-vis a (postulated) third factor (the divine, the truth, ultimate reality). This is the decisive criterion in most external definitions of religions as cultural interpretive systems *and* the decisive factor in the internal self-understanding of the relig-

ions, also in the light of the environmental crisis. It is the (intricate and problematic) relations of 'nature' and 'the supernatural' that matter and enable the religions to present themselves as alternative or supplementary systems of interpretation, ethics and action.
- In all the religions popular ideas and practices, due to the historical influences of earlier cultures and religions and the continuous presence of believers whose religious aims are related to an economy based on direct relations to the near- environment, coexist with the theological systems.
- In all the religions traditions of mysticism influence the discourses on the divine, man and nature. They contribute to bridging the opposites of the dualistic schemes otherwise separating the transcendent divine from the immanent man and nature. Ascetic ideals and practices furthermore function as models for a non-materialistic, non-consumerist attitude to life.

A critical assessment

Trying to combine scientific, ethical-political and environmental concerns is no easy task, and I have no intention to realize this ideal in these few concluding remarks. All I want to do is to point to some areas in need of more research and to some – in my opinion – positive as well as negative aspects of the combination of religion and conservation.

Thanks to scholars of religion and anthropologists, it is by now a well-established fact that the popular notions of 'nature-religions', 'nature-worship' and the 'ecological wisdom' of primitive societies are loaded with problems. I think, however, that science as well as the environment could benefit from further studies in the 'myth'. Not as an unscientific 'false story', but as true myth 'lived' by the believers, be they indigenous peoples themselves or New Age – environmentalists. A combination of an academic interest in contemporary religion with a concern for conservation calls for more empirical studies evaluating the environmental consequenses of the applications of the myth. Likewise, studies in the relations between other kinds of religiosity and environmentalism ought to be initiated. The environmentalist who, for instance, claims cosmic and mystic 'peak-experiences' standing on top of a computer-steered heap of compost, is an obvious subject for the scholar of religion interested in contemporary forms of religious activity. He is, however, also evident subject-matter for those who want to see under what circumstances religion and environmentalism can become partners and not enemies. And, of course, more critical analyses of the possible neo-imperialist and neo-colonialist implications and political use and abuse of the myth, in the West and elsewhere, are also needed.

Turning to the alliance of the world religions and conservation, I find it safe first to say that the analysis of the greening of the religions demonstrates the validity of the general theory of religion put forward here. The development makes it possible to look directly into the workshops where religions and reinterpretations are created and recreated, where traditions are reinvented. Modern ideas about 'nature' and 'ecology' are read back into classical texts and used to prove the eternal wisdom of the gods, the eternal validity of the religion and it's ethics, and to develop environmentally sound thinking and procedures. Classical key-concepts get new or extended meanings and applications, but the modern procedures of reinterpretation are often in perfect accordance with traditional procedures for adjusting the religious traditions to new historical situations. Reading discourses on 'nature' or 'pollution' into Buddhist or Muslim classical sources differs in principle not from classical Christian reinterpretations of the Old Testament. The prefigurations of ecological worldviews resemble the prefiguration of Christ!

As for the cross-cultural and syncretistic aspect of the development, one can see how the religions or denominations within them, for instance within Christianity and Buddhism, fertilize each other across doctrinal and cultural frontiers. British Buddhism and British Hinduism interact with Thai-Buddhism and Hinduism in India or Bali, and European or American sufi-inspired Muslims inspire orthodox Muslims elsewhere. The work done by various Christian organizations, especially the World Council of Churches, reveals the universal, multicultural and cross-cultural nature of the process and of modern Christianity: local theologies, like the Waterbuffalo-theology related to an Asian culture and environment or the Coconut-theology growing on the islands in the Pacific, cross-cultural and theological borders to the West. As New Age theology inspired by ideas about primitive religion crosses the Atlantic and finds its way to European Christianity, so do these theologies. Ecological theology is ecumenical, cross-cultural and inter-religious.

Stating that the greening of the religions is quite natural from the point of view of the history of religions and good for studies in religion and conservation, is one thing. Whether it is 'good' for persons not directly involved by their own free will, for the environment, for the various local communities and for the world community is quite another question, which I have no intention to deal with in an exhaustive way. Only this much can be said:

I find the reasoning behind the WFF project fairly sound and balanced. The participation of the world religions in environmental projects does, in the light of the dimensions of the crisis and from a pragmatic point of view, add a substantial and needed amount of manpower to global and local projects. I also find it useful that representatives from various cultures, or various interpretive communities (the religions, the communities of scientists, economists, environmentalists, etc.), cooperate in a way that can lead to mutual inspiration and a balancing of opposite views and various means. To return to Mt. Everest: since Mt. Everest in the time before and after it was conquered evidently was not protected (at least not very well if judged from the present state of affairs) either by Buddhist or materialistic spirits and notions, one may of course say that the general discrepancy between worldviews and practice, be they religious or not, is once again proved. Admitting the relevance and relative truth of such an argument does not, however, exclude the possibility that religiously based concepts of nature and a religiously based environmental ethics may be of some help, nor that the combination of religious and non-religious ideas and means may be useful. The cooperation, so to say, of Edmund Hillary and Sherpa Tenzing, may work better than no cooperation. If biological diversity is good for the environment, maybe conceptual and practical, cultural and religious diversity and interaction, is good too.[14]

Naturally, some Buddhists, Hindus or Christians may entertain religious ideas and sentiments, for instance about the godgiven 'naturally natural' or the 'pure and impure', that run contrary to scientifically, well proven and effective methods, means and materials. The analysis of the factual projects within the religions, however, does not point in that direction, and the leading figures in the involved groups do not believe that faith alone can move mountains. In general, the greening of the religions within the WWF project cannot be interpreted as an irrational return to some 'supernatural' belief in 'faith healing'. It is, moreover, frequently based upon clear notions of the former failure of the religions to effectuate their postulated environmentally benign attitudes and views, and upon a recognition of the fact that not all believers are practicing believers. Neither in matters of religion in general, nor in matters concerning the environment. The work within the alliance is to a large degree based upon a balanced hope that faithfulness to certain religious beliefs, combined with pragmatic, rational and scientifically thinking, may be conducive to more environmentally benign thinking and acting. The religions and the adherers certainly will not be able to make mountains move. They may, however, contribute to efforts trying to stop mountains from turning into heaps of garbage.

Other considerations of other, not so benign, consequences of the environmentalist movements within the religions naturally must also be voiced. The risk of having religious authorities play an elitist game in international and environmentalist politics and ethics running counter to the interests of local people, is obvious. Also, one must naturally be aware that adoption of international perceptions and concepts of nature may contribute more to the well-being of some "pieces of nature" and to an overall and artificial unity than to cultural diversity. Even if the participants in the WFF-projects deliberately try to accommodate the messages and means to local religious traditions and celebrate the "richness of the diverse religious traditions" through traditional rituals, festivals and symbols.

Finally, of course, everybody has to decide for himself to what a degree he thinks that societies and international organizations defining themselves as secular find the increasing religious activities a desirable 'good'. The greening of the religions can, as mentioned earlier, be seen as one of the ways by which religions reenter the scenes where global ethical (and therefore also political) issues are discussed and negotiated.[15] Is this a contribution to the democratic and open-ended process of, hitherto mostly secular, negotiations? Is the introduction of the religious worldviews just adding to the celebrated plurality, or do they – in the long run – represent a threat to pluralism, democracy and a rational, secular discourse? In spite of the fact that religions change, it is still a central part of their self-understanding that they continue to exist and accommodate exactly because they represent an eternal truth, the true way of viewing the world and deciding what is god-given, natural, true and right. In relations between men, and in the relation between man and nature.

Ethics and politics cannot be separated. The religious environmental ethics is part of the continuous struggle for the right and power to decide and define what counts as real, good, normal and natural. What is worth conserving and what is not. Viewed in this perspective, one may very well invest some hope in the religions as partners in environmental efforts, but one may equally well fear them.

Notes

1   The amount of source material, i.e. publications by various religious environmentalists, is stunning. An anthology displaying the spectre is Gottlieb, R.S. (ed.): *This Sacred Earth. Religion, Nature, Environment,* New York-London: Routledge 1996. The anthropological perspectives are eminently set forth and discussed by Milton, K.: *Environmentalism and Cultural Theory,* London-New York: Routledge 1996. For an excellent survey of 'primitivism' see Geertz, A.W. 'Can We Move Beyond Primitivism? On Recovering the Indigenes of Indigenous Religions in the Academic Study of Religion', in: Olupona, J. (ed.): *Beyond 'Primitivism': Indigenous Religions and Modernity* (in print).

2   See Jensen, T.: 'Religionerne og religionshistorikeren i naturkampen', *CHAOS,* Dansk-norsk tidsskrift for religionshistoriske studier nr. 14, 1990, pp. 32-51, and Jensen, T. and M. Rothstein: *Gud og grønne skove. Religioner og naturbevarelse,* København: Munksgaard 1991. Tucker, M.E.: 'The Emerging Alliance of Religion and Ecology', *Worldviews,* vol. 1, number 1, April 1997, 3-23 is the first one to survey some part of the same area with a focus on North American alliances. Kinsley, D.: *Ecology and Religion:*

Ecological Spirituality in Cross-Cultural Perspective, Prentice Hall 1995, may be added, but the critical dimension is lacking. A more critical survey is given by Pye, M. et al.: 'Ökologie und Religionen', *Marburg Journal of Religion,* Volume 2, nr. 1, 1997, 4 pages, 4782 words. See also Harris, I.: 'Buddhist Environmental Ethics and Detraditionalization: The Case of EcoBuddhism', *Religion* 25, 1995, pp. 199-211. Publications by authors affiliated to the WWF project, either as consultants or as members of participating religious groups, mostly represent a kind of mixture between dedicated scholarly presentations and source material.

3   Older sources are presented in Jensen & Rothstein, *op. cit.* Later sources are mainly the materials produced by the WWF: *The New Road: The Bulletin of the WWF Network on Conservation and Religion,* Gland, Switzerland; Edwards, J. and M. Palmer (eds.): *Holy Ground. The Guide to Faith and Ecology,* Pilkington Press 1997; the full revised declarations by the participating nine religions and sheets of information issued by the WWF U.K. and ARC; the series on *World Religions and Ecology* by Cassell Publishers in cooperation with the WWF.

4   Jensen, T.: ' Gud mod år 2000', in: Müller, M. (red.): *Politikens bog om religioner og religiøse bevægelser,* København: Politikens Forlag 1996, pp. 8-57, and Jensen T: 'Familiar and Unfamiliar Challenges to the Study and Teaching of Religions' in: Holm, N.G. (ed.): *The Familiar and the Unfamiliar in the World Religions: Challenges for Religious Education Today,* Åbo: The Academy of Åbo Press 1997 pp. 199-220.

5   Geertz, A.W.: *The Invention of Prophecy. Continuity and Meaning in Hopi Indian Religion,* Aarhus: Brunbakke Publications 1992, p. 21. An introduction to contemporary discussions on the theories and definitions of religion is found in McCutcheon, R.T.: 'The Category ' Religion ' in Recent Publications. A Critical Survey', *Numen,* vol. XLII, no 3, october 1995, pp. 284-309. See also Bilde, P.: 'Begrebet Religion' *CHAOS* nr. 15, 1991, pp. 3-25 and Geertz, A.W., ' Begrebet religion endnu engang – et deduktivt forsøg', *CHAOS* nr. 26, pp. 109-128.

6   For contemporary studies in tradition and renewal, see Geertz, A.W. & Sinding Jensen, J. (eds.): *Religion, Tradition and Renewal,* Aarhus: Aarhus University Press 1991.

7   Cf. Pedersen, P.: ' Nature, Religion and Cultural Identity: The Religious Environmentalist Paradigm', in: Bruun and Kalland: *Asian Perceptions of Nature. A Critical Approach,* Richmond: Curzon Press, pp. 258-276.

8   Edwards, J. and M. Palmer (eds.): *op.cit.* p. 36.

9   Edwards, J. and M. Palmer (eds.): *op.cit.* p. 43.

10 For a comprehensive description and analysis of the Assisi-event cf. Jensen & Rothstein *op. cit.* The Assisi and Windsor declarations are published in Edwards, J. and M. Palmer (eds.) *op.cit.* The declarations issued at the festival on Religion and Conservation in Copenhagen are published in Jensen, T. (ed.): *Religioner og naturbevarelse,* København: WWF Verdensnaturfonden 1990. Some of the present projects are presented in Edward, J. and M. Palmer (eds.): *op.cit.*

11 *News From ARC,* Bulletin of the Alliance of Religions and Conservation, Issue no.1, Spring 1997 p.3.

12 Some of the mentioned projects are relatively new and described in recent press-releases from ARC. Other examples can be found in Jensen & Rothstein: *op.cit.*

13 A complete description of the concepts can be found in Jensen & Rothstein: *op.cit.* and in Jensen, T.: 'Religions and the Environment in the Classroom', in Doble, P. and M.

Hayward (eds.): *The Contribution of Religious Education to Teaching about the Environment. Report of a Conference held at York, UK, September 1992,* York: York Religious Education Centre Univ. College of Ripon and York St. John 1993, pp. 72-89.

14 This does not mean that I do not see some scholarly and practical problems connected to e.g. some concepts of 'culture' implied in the projects. Cf. Milton, *op. cit.* p. 198f.

15 On this, see the articles refered to in note 6.

Policy discourses on environmental problems in Ecuador and Norway – A comparative perspective

Randi Kaarhus

The natural sciences have provided us with a basis for defining environmental problems in *universal* terms. Thus, in the world of the present, it is possible to conceive of environmental problems as cross-cultural and common *global* problems. At the same time, environmental problems are inextricably intertwined with the multiple relationships of human societies with *nature,* as these problems manifest themselves in a diversity of localized social contexts. Furthermore, at different levels from the local to the global, environmental problems are actually subject to continuous processes of political negotiation and cultural redefinition.

The present article will focus on how what may be seen as the *same class* of environmental problems are constructed as *different* problems in distinct social and political contexts.[1] Drawing upon the universal terms of natural-science knowledge, certain outcomes of the complex processes involved in *soil erosion* can be defined as such a *class of problems*. One might say that, basically, soil erosion is a natural phenomenon. *Increasing rates* of soil erosion are, however, often the unintended consequence of human activities such as agricultural production. In many cases, agriculture results in the degradation of soils; soil particles and plant nutrients being transported away from cultivated fields by wind and running water as "*erosive agents*".[2]

With reference to soil erosion as a global or 'universal' class of problems, the present article will describe *how* the localized outcomes of processes of soil erosion have been *constructed* as environmental problems at the national level in two highly distinct cultural and socio-political settings – more specifically, in the policy and planning *discourses* formulated in the modern nation states of Norway and Ecuador. In the Norwegian context, the outcomes of such processes of soil erosion have, in fact, almost exclusively been defined and discussed in terms of problems of *pollution* – water pollution. In the Ecuadorean context, the outcomes of "the same" processes have, by contrast, almost exclusively been defined and discussed in terms of *land degradation*.

One might ask, in what sense do such distinct problem constructions actually refer to "the same" problem? This question, in turn, brings up the more general question of *comparability*. Let me start with a brief discussion of the possibilities of comparison in the field of enquiry constituted by environmental *problems* and problem *constructions*.

Comparison and comparability

Comparison may be seen as quite basic to human thought. Conceptions of *likeness* and *difference* will, at some level, imply comparison.[3] Comparison will also be central to processes of *classification* – including scientific classification. And classification will necessarily involve making phenomena and processes comparable by means of identifying similarities and differences. To identify phenomena or processes in the real world as members of a *class* will, in some way or other, involve characterizing the members of this class as instances of "the same".

Let me exemplify this preliminary argument with a brief reference to a typically *global* environmental problem, a problem which at the international arena is labelled "the emissions of substances that deplete the ozone layer". Local emissions of substances that deplete the ozone layer are generally considered to be specific instances of one and the same global problem, which, in turn, is made up by the total emissions of such substances. The relationship between the local and the global will in this case simply be one of aggregation; the global problem being made up by the aggregate of all local emissions.

This global environmental problem – a problem which can only be identified and defined in universal science-based terms – can, on its part, be juxtaposed to the typically *localized* problems of soil erosion on cultivable land. It may, in fact, be claimed that the principal characteristic of problems of soil erosion resulting from agricultural production in, say, countries as different as Norway and Ecuador, is precisely their distinctiveness. One might even sustain that the particular constellation of geographical, historical, social, cultural, and political factors in each case requires a description of the unique characteristics of the case in its particular context.

But if each case is unique, what about comparability? Following Leo Howe, I would sustain that neither uniqueness nor comparability are necessarily given in the empirical phenomena themselves. According to Howe *"uniqueness is not an inherent attribute of phenomena, whether these be natural or social. If something is unique it is so because we choose to classify it in terms of criteria which differentiate it from everything else; we could just as well select other criteria which would classify it as one of a kind. In short, the paradox which states that the comparison of similar phenomena leads to triviality while the comparison of different phenomena is impossible can be seen for the confusion it is."*[4]

I would say that *comparability* does not simply depend on likeness or difference as inherent properties of the phenomena compared. In studies of social and cultural phenomena, comparability will to a certain extent be *constructed* through the process of comparison itself. But what form of comparison are we then talking about? Theda Skocpol and Margaret Somers have suggested that the *contrast of contexts* is one possible *logic of comparison* which can be used both in history and social science. This logic of contrasting contexts can serve precisely to *"bring out the unique features of each particular case ... and to show how these unique features affect the working-out of putatively general social processes."*[5]

With regard to contemporary environmental problems, there does, however, exist a putatively universal language of description. This is a language based on current knowledge in the natural sciences. The existence of such a language is a social fact, though the context-independent *objectivity* of the knowledge on which it is based has been an issue of renewed philosophical debates.[6] With regard to environmental problems, there is, however, considerable agreement regarding the crucial role of science-based knowledge, both in the identification and representation of environmental problems as real phenomena in the world of the present and as problems which must be taken seriously as policy issues at both national, regional, and global levels of contemporary policy making.[7]

What I would like to point out here is that the actual use of such a science-based language as a *common repertoire* drawn upon in the formulation of diverse localized discourses provides interesting possibilities for comparative sociocultural analysis. Let me develop some more on this point by briefly introducing one such repertoire of science-based representations. In this case, the most relevant repertoire will be the one drawn upon in science-based descriptions of that class of environmental problems which result from the combination of certain human agricultural activities and the 'universal' processes of soil erosion.

Science-based representations of soil erosion

At present there exists – to my knowledge – one comprehensive science-based assessment of soil erosion as an environmental problem at a global scale. It was published in 1990 in the form of a *World Map*, which covers the status of the soils on most of the surface of the earth. According to figures based on this World-Map material, about 15% of the total land surface of the earth is affected by what is called "*human-induced soil degradation*". If we turn to particular regions, we find that in Europe as much as 23% of the total land area is affected, while about 14% of the land area of the South-American continent is affected by similar processes of human-induced soil degradation.[8]

Now, what sort of knowledge forms the basis for making these assessments? Or: *how* do we know *what* we know about soil erosion as an environmental problem – from the local to the global scale? In very general terms, one could say that current science-based representations of soil erosion are based on two distinguishable *categories* of knowledge; each category drawing upon several scientific disciplines.

On the one hand, there is the *detailed knowledge of soils* as such. This category comprises the science-based knowledge of the particles which constitute *the soil*, this very basic – and highly fascinating – substance which covers most of the earth's surface. A large number of natural processes interact in the formation and reproduction of this substance which is positioned at the *interface*, one might say, of the living and the non-living. The category of knowledge we refer to here includes science-based representations of the multiple biotic and abiotic processes interacting in the continual generation and regeneration of different types of soils. These are soils which, in turn, can be classified by drawing upon diverse criteria of classification. Furthermore, we refer to knowledge on the processes by which different types of soils are affected by *agents of soil erosion* such as water and rainfall, snowmelt and wind. These 'agents', on their part, act together with the continual pull exercised by the *natural law* of Newtonian gravitation to produce what we call soil erosion.[9]

The multiple chemical, mechanical, and biological processes involved in the processes of soil erosion do not, however, lend themselves very easily to measurement and quantification. The measurement of soil erosion in the field is, for instance, an extremely elusive task. But in this case, like in a number of other cases, the fact that the results of *measurements* with regard to environmental problems are fraught with uncertainty, does not mean that the problems themselves are not *real*.

There is, however, also another *category* of science-based knowledge dealing with the processes and substances involved in soil erosion which is relevant here. It is a category which provides a somewhat different perspective on the processes involved, even though both perspectives draw upon a common repertoire of science-based knowledge and must be seen as complementary, rather than contradictory. This second category of knowledge is primarily concerned with the cycling of matter in so-called *biogeochemical cycles*. The cycling of elements such as carbon, nitrogen, phosphorus, and sulphur have been modelled – and to some extent measured – at local, as well as global levels. This category of science-based knowledge – and the uncertainty involved – has received a great deal of public attention over the last years, especially in connection with the risks of global warming resulting from increasing emissions of carbon dioxide.

According to existing science-based knowledge on this cycling of elements, it is, however, *not* the carbon cycle which has been most substantially altered by human activities. Human activities have interfered even more with the *nitrogen cycle*. The major factor in this alteration is, in fact, agriculture; modern agriculture being, to a great extent, based on the production and use of chemical nitrogen fertilizers.[10]

One might also say that modern agriculture consists in harnessing elements such as nitrogen and phosphorus in plant growth; that is, harnessing these elements in the production of food for a growing human population – and for a growing global market. On the other hand, processes of soil erosion result in the same nutrients being carried off from cultivated – and fertilized – soils to watercourses and sea areas. Here both nitrogen and phosphorus will serve as nutrients for freshwater as well as marine organisms. This erosion of nutrient elements from agricultural lands to watercourses and sea areas is usually called *run-off*.

While the increased cycling of carbon to the atmosphere apparently will result in some degree of global warming, the increased cycling of nitrogen and phosphorus to water has in many cases resulted in *eutrophication* – i.e. excessive algae growth or what is sometimes called "algae catastrophes". These problems have received a great deal of attention in the North Sea area, and I will return to this regional problem in the discussion of the Norwegian case below.

To sum up my presentation so far: I have defined *soil erosion* as a class of environmental problems with reference to the multiple natural and human-induced erosive processes that affect soils. The processes are *natural* in the sense that they will continue without human intervention, though they are often accelerated through human activities – in particular, agriculture. Our knowledge of these processes is to a great extent based on available knowledge produced in the *natural sciences*. But we also know that the actual manifestations of these problems and their effects are deeply interlinked with social, political, and cultural processes in different parts of the world.

Now, one may ask: how are these interlinked processes *represented* in the authoritative discourses produced in distinct socio-political contexts? In my view, a careful juxtaposition of 'contrasting contexts', a juxtaposition which also indicates how a common repertoire of representations is drawn upon and interpreted in distinct settings, may serve to illuminate several aspects of the complex relationships of *human societies* with *nature* which are intrinsic to present-day environmental problems. This form of comparative analysis might also heighten our awareness, not only of the socio-cultural conditioning, but also of the political processes involved in the construction of environmental problems as *real problems* in different parts of our common world.

Soil erosion as an environmental problem in Ecuador

I have already given one clue to the construction of soil erosion as an environmental problem in Ecuador, observing that processes of soil erosion primarily have been treated in terms of *land degradation* in the context of Ecuadorean policy discourses. At this point, I would add that land degradation to a great extent has been associated with what has been called "*the pressure of people on land*". This association was made explicit already in the first policy document on agricultural development in Ecuadorean policy making, the *Programme of agricultural development* of 1964.[11]

In this policy document, two major social problems considered to be of fundamental importance at the time, were also associated with land degradation. One was the *unequal distribution* of land, especially in the Andean highlands of Ecuador. The other was called "*lack of knowledge*", and referred more particularly to a lacking knowledge of modern methods in agriculture and land management on the part of the highland Indian population. Land degradation was, in this context, not seen as a 'modern' environmental problem, but rather as a historical product and a result of backwardness and underdevelopment. The *Programme of agricultural development* synthesizes these interlinked problem conceptions in a single statement: "*In the highlands, the balance between man and land entails a heavy pressure of man on a land which is badly distributed and exhausted after centuries of deficient management.*"[12]

In the early 1960s, the traditional *hacienda* system, which was characterized by a small group of powerful landowners and a large population of subjugated Indians, still dominated the Ecuadorean highlands. But in Ecuador, as in the rest of Latin America, the 1960s witnessed a growing pressure for land reforms. In fact, control of land constituted the major political issue of the time. While the traditional elite struggled to maintain their historical control of *people* through control of *land*, other groups promoted the ideas of land reform and redistribution. In this context, the problem of *land degradation* became one of the arguments in favour of land reform.

Now, one could say that Malthusian conceptions of population-pressure on agricultural land as an inevitable outcome of population growth have been influential, not only in the social sciences of economics and demography, but also in modern planning and policy making, as well as in the more recent discourses of modern environmentalism. But this *"pressure of people on land"* is definitely hard to handle as a political problem. Since any large-scale *redistribution of land* also involves a redistribution of political power, radical land reforms have proved very difficult to carry through, not least in Latin America.[13]

Nevertheless, several programmes of agrarian reform were planned and – in part – carried out in Ecuador in the 1960s and 1970s. In practice, they also involved a certain redistribution of land in the Andean highland region. But a central element in the land reforms actually consisted in programmes aimed at 'colonizing' the Ecuadorean part of the Amazonian rainforests. The objective was to resettle parts of the highland agricultural population *"which grows more every day"* in the *"vast reserves of agricultural territory"* of the lowland areas.[14]

The environmental problems resulting from this relatively recent intra-state 'colonization' of the Amazonian rainforests in Ecuador, as well as in other South-American countries, have attracted a good deal of attention on the international scene. But within Ecuador, the implementation of the land reforms is considered to have resulted in equally serious environmental problems in the highland region – in the form of soil erosion. To a great extent, the outcome of the redistribution of agricultural lands was that the best lands in the valley bottoms remained in the hands of the former *hacienda* owners, while the marginal lands on the mountain slopes were distributed or sold to the former Indian tenants and other small-scale farmers. The large farms in the valley bottoms mostly turned their fertile lands into pasture for dairy cattle, while the former tenants on the small plots on the sloping hillsides would dedicate their lands to the intensive cultivation of foodcrops.[15]

This pattern of land use has resulted in accelerating problems of soil erosion, above all on the marginal and sloping lands where the former tenants have settled to cultivate their allotted portions of land. These problems are also recognized in Ecuadorean policy discourses. In the *National plan of economic and social development* for the period 1989-1992, soil erosion in the Andean highlands was, together with deforestation in the Amazonian lowlands, pointed out as the most serious environmental problem in present-day Ecuador. Here it is stated that: *"Great extensions of land in the Ecuadorean highlands are eroded, in some cases at a really critical level."*[16]

In the context of Ecuadorean policy making, soil erosion is presented as a serious problem because it involves the degradation of agricultural land, not only as a natural resource, but also as economic, political, and cultural *capital*. We can now see that the land reforms of the 1960s and 1970s failed to solve the problems of erosion which were identified in the early 1960s. The problems were rather aggravated as an unintended consequence of the – admittedly partial – implementation of the land reforms.

Now, my intention here is neither to criticize the land reforms as such, nor to point out any final solution to the problems of soil erosion in the Ecuadorean highlands. What I have wanted to show is *how* soil erosion, as a particular class of environmental problems in natural-science terms, in the "real world" is interlinked with historical processes, social conditions, and political action. I hope I may further illustrate this point by turning to my contrasting case of Norwegian environmental-problem construction.

Soil erosion as an environmental problem in Norway

In the Norwegian context, soil erosion was considered to be a non-existent problem in the 1960s and -70s. Then, in the late 1980s, all of a sudden it was discovered that processes of soil erosion were also affecting Norwegian soils. This discovery was – at least in part – the result of a series of regional agreements to protect the North Sea. In 1987, eight North Sea countries signed the so-called London Declaration, agreeing to reduce emissions of phosphorous and nitrogen to vulnerable areas of the North Sea by 50% between 1985 and 1995. The sources of these emissions included industry, municipal waste water, and agriculture.[17]

A few months later, the signing of this North Sea Declaration was followed by a sudden bloom of a marine algae – *Chrysochromulina polylepsis*. In the spring of 1988, this bloom raised great alarm along the southeastern coast of Norway. The algae produced a toxic substance which killed off other marine organisms, and dead fish were observed floating along the coast. Biological research concluded that the "algae catastrophe" was a result of unusually high concentrations of nitrogen in the sea. This was, in turn, taken as a warning of the urgency to follow up the London Declaration, by defining concrete means of reducing the nutrient emissions to the North Sea.

In the Norwegian case, it was primarily as a result of the London Declaration, in combination with the algae catastrophe, that processes of erosion on agricultural lands came to be constituted as a research issue in several scientific disciplines. In this situation, there was suddenly a demand – on the part of the Norwegian state – for science-based knowledge on the cycling of nitrogen and phosphorus from land to water. This category of science-based knowledge was used as an input to *linear programming models*, which were constructed to represent the processes of *nutrient runoff* from cultivated fields through subsurface water and watercourses further on to the North Sea. Such models were to play a central role in the national follow-up of the North Sea Declarations. Both scientists and state authorities were involved in these efforts to construct a knowledge base for the so-called *monitoring*, both of the nutrient runoff and of the expected effects of the measures taken to reduce emissions.

The London Declaration had placed nutrient runoff from agriculture within a general framework of problems of *pollution*. This conceptualization of erosive processes within a framework of pollution corresponded to the dominant perspective on environmental problems in the most influential scientific discipline in Norwegian planning and policy making from the 1950s onwards: economics.[18] Environmental economics in Norway was primarily concentrating on the construction of models representing *"emissions"* and *"pollution"*.[19] Knowledge of the biogeochemical cycling of nutrient elements and economic models of pollution were now brought together in order to design policy measures that would reduce emissions of nitrogen and phosphorus from agricultural lands to the North Sea.

A model which represents cultivated land as a source of polluting emissions in many ways contrasted with other – and more common – conceptions of agriculture. Even in highly modernized agriculture, the biological processes involved in plant growth do not necessarily fit into simple models of pollution taken from industrial production. Yet, it must be admitted that modernized high-input agriculture increasingly has come to acquire many of the characteristics of a highly specialized industrial production.

In this context, it must be noted that agricultural production in Norway was thoroughly transformed during the period from the 1950s to the 1980s. This transformation resulted in a regional specialization in production, with grain production being concentrated to the lowlands in the south-east. This "channelling" of grain production to the southeast, and the corresponding "channelling" of milk production to more marginal zones have actually constituted a basic element in Norwegian agricultural policies from the 1950s onwards. The *channelling policy* has also been the foundation of the political consensus which has characterized agricultural politics in Norway during most of the post-war period.[20] Not until the end of the 1980s did Norwegian authorities recognize that this very basic element in modern Norwegian agriculture had also resulted in serious environmental problems.

The problems of nutrient emissions were, to a great extent, the result of the high levels of agrochemical *inputs*, in particular nitrogen fertilizer which is used in specialized grain production. Part of the nitrogen contents of chemical fertilizers are taken up by growing plants. But with high levels of fertilizer application, considerable proportions of nutrient elements inevitably *run-off* as a result of the erosive processes that continually affect cultivated soils. When these processes were suddenly brought up on the political agenda in the late 1980s, it was a result of the fact that the zones of specialized grain production in Norway precisely are those which drain into what had been defined as *"the most vulnerable areas of the North Sea"*.[21]

It might be asked: what is the role of the *soil* as a *natural resource* in this modern and highly specialized form of agriculture? One could answer that the role of the soil has primarily become one of keeping the growing plants in place while, at the same time, transmitting agrochemical inputs to plant growth. In such a perspective, the soils of cultivated lands constitute, not so much a *natural resource* as a *place of transit* for the cycling of inputs and outputs; *inputs* to crops and *outputs* of crops – and, we may add, runoff of excess nutrients.

Concluding remarks

Let me conclude this juxtaposition of two distinct cases of environmental-problem construction with a few more general remarks. One may say that it is possible to define environmental problems at local and national levels as instances of *the same* class of – global – problems with reference to a common framework based on natural-science knowledge. On the other hand, environmental problems have to do with the complex relationships of *human societies* with *nature*; they have to do with the use and exploitation of material resources, as well as with the conceptualization and use of symbolic constructs such as the *nature/culture* divide. And at the level of contemporary nation states, a number of particular historical, socio-economic, political, and cultural conditions will in practice structure what we may call authoritative *policy discourses* concerning environmental problems.

In this article, I have shown that in Ecuador soil erosion has for several decades been constructed as a problem of *land degradation* – or *resource degradation*; land being a basic natural resource both in political, social, and cultural terms. In Norway, by contrast, I have argued that the problems resulting from processes of soil erosion have, since the 1980s, primarily been conceived in terms of *pollution*. Elsewhere, I have given a more comprehensive account of the contexts of these different problem constructions.[22] Here I will just point out some implications as regards *comparative studies* in the field of environmental problems and problem constructions.

In my view, a comparative *contrast of contexts* – such as the one briefly delineated here – may not only serve to illuminate what may be seen as *"one set"* of less-understood phenomena *"by reference to another set more clearly comprehended"*;[23] or to describe a less familiar setting – in this context Ecuador – by reference to a more familiar setting – in this context Norway. I would also suggest that the exploration of a case constituted by a less-known set of phenomena provides a position from which a more familiar set can be *re-illuminated* – and, possibly, interpreted in a new perspective. This form of comparison provides a point of departure for an exploratory movement back and forth between the more-familiar and the less-familiar contexts and sets of phenomena. This is, in turn, a process which may open up for a more *self-reflexive construction* of similarities and differences between cases.

I also think that this form of comparative analysis may be drawn upon in efforts of cooperation across national and cultural borders. If we accept that environmental problems are *both* objective phenomena and social constructions, we should, in principle, also see the possibility of *diverse* problem constructions – produced in different settings – being brought together to serve as an extended knowledge base in cross-cultural efforts to protect nature and the human environment.

Notes

1. The article is based on research funded by a grant from the Research Council of Norway and additional economic support from the Norwegian Institute for Urban and Regional Research (NIBR). I would also like to thank Nora Gotaas and Ingeborg Svennevig for their comments on the first draft of this article.
2. Morgan, R.P.C.: *Soil erosion and conservation*, Harlow, Essex: Longman 1986, p. 12.
3. Cf. Holy, L. (ed.): *Comparative anthropology*, Oxford: Basil Blackwell 1987.
4. Howe, L.: "Caste in Bali and India: levels of comparison", in L. Holy (ed.): *Comparative anthropology*, Oxford: Basil Blackwell 1987, p. 136.
5. Skocpol, T. and M. Somers: "The uses of comparative history in macrosocial inquiry", *Comparative Studies in Society and History*, Vol. 22, 1980, p. 178.
6. These debates will not be followed up here, but see e.g. Bernstein, R.J.: *Beyond objectivism and relativism*, Oxford: Basil Blackwell 1983; Giere, R.N.: *Explaining science: a cognitive approach*, Chicago: University of Chicago Press, 1988.
7. Cf. Worster, D.E.: *Nature's economy: A history of ecological ideas*, Cambridge: Cambridge University Press 1994; Pepper, D.: *Modern environmentalism: An introduction*, London: Routledge 1996.
8. *World map on status of human-induced soil degradation*, Nairobi: UNEP/ISRIC 1990; Oldeman, L.R., R.T. Hakkeling and W.G. Sombroek: *World map of the status of human-induced soil degradation: an explanatory note*, Wageningen: ISRIC/UNEP.
9. Cf. Kaarhus, R.: *Conceiving environmental problems: A comparative study of scientific knowledge constructions and policy discourses in Ecuador and Norway*, Ph.D. thesis, University of Oslo 1996. NIBRs reprint 20/1996, pp. 84-96.
10. Stewart, J.W. et al.: "Global cycles", in: J.C. Dooge et al. (eds.): *An agenda of science for*

environment and development into the 21st century. Cambridge: Cambridge University Press 1992, p. 134.
11 JUNAPLA: *Programa de desarrollo agropecuario*, Quito 1964. JUNAPLA was the acronym of the *Junta Nacional de Planificación y Coordinación Económica*, the Ecuadorean National Board for Economic Planning and Coordination.
12 JUNAPLA 1964, p. 24. My translation from the original Spanish text.
13 Cf. Dorner, P.: *Latin American land reforms in theory and practice: A retrospective analysis*, Madison: University of Wisconsin Press 1992.
14 JUNAPLA 1964, p. 30. My translations from the original text in Spanish.
15 Cf. De Noni, G., M. Viennot and G. Trujillo: "Soil erosion and conservation research in Ecuador", in: H. Hurni and K. Tato (eds.): *Erosion, conservation, and small-scale farming*, Bern: Geographica Bernensia 1992.
16 Gobierno Constitucional de Ecuador: *Plan nacional de desarrollo económico y social 1989-1992*. Quito: CONADE 1989, Vol. III, p. 632.
17 Cf. Report to the Storting No. 64: *Concerning Norway's implementation of the North Sea Declarations*, Oslo: Miljøverndepartementet 1991-92, p. 17.
18 Cf. Kaarhus *op.cit.* pp. 167-199.
19 Cf. Førsund, F.R. and S. Strøm: *Miljø-økonomi*, Oslo: Universitetsforlaget 1994 [1980].
20 Cf. Njøs, A.: "Soil erosion: a problem for agriculture and environment in Norway", in H. Lilleng and B. Rognerud (eds.): *Environmental challenges and solutions in agricultural engineering*, Ås, Norway: Norwegian Commission of CIGR/ Agricultural University of Norway/ Jordforsk 1991; Vatn, A.: "Norsk landbruk – frå Hitra til GATT", in: J.W. Simonsen and A. Vatn (eds.): *Landbruk i endring*, Oslo: Universitetsforlaget 1992.
21 Report to the Storting No. 64: *Concerning Norway's implementation of the North Sea Declarations*, Oslo: Miljøverndepartementet 1991-92.
22 Kaarhus, R. *op.cit.*
23 Howe, L.: "Caste in Bali and India: levels of comparison", in L. Holy (ed.): *Comparative anthropology*, Oxford: Basil Blackwell 1987, p. 136.

Trade offs in joint implementation strategies

The Central American forestry case

Klaus Lindegaard and Olman Segura

Joint Implementation

The traditional strategy towards international environmental action is mainly based on two different programmes. First, the special programmes of the international organizations supporting projects in the poor countries with the financial means allocated from the richer countries, or second, on international agreements between problem-generating countries, which are implemented nationally sometimes with some special support schemes or arrangements for the poorer countries. The policy of joint implementation (JI) is emerging as a new strategy for implementing global environmental aims, especially regulating the climate change process, where funds from rich problem-generating countries are allocated directly for projects in other (poor) countries based upon a mixed argument of partnership and cost-effectiveness.

In principle, there are different types of projects, either allocating funds for the solution of environmental problems in poorer countries or allocating funds for remedial or compensatory action against ones own problems in poorer countries. The paper reviews the divergent views on the benefits and future use of bi-lateral national agreements of JI on the basis of the experiences gained so far with Central American forestry projects. Here the conflicting views of national governments and local community and environmental groups address whether JI projects are functioning as global partnership and development aid or as indulgence for business-as-usual practices and eco-colonialism. The local-local international partnerships strategy (e.g. NGOs and business industry projects of JI schemes in the Central American forestry sector) which are often complicated by the same opposing arguments, are included in the assessment and discussion of the possible future implementation of global environmental partnership strategies.

The absence of supranational authorities with the power to implement the polluter pays principle (PPP) through pollution taxes and marketable pollution permits together with the general lack of economic incentives in the international allocation of abatement effort, has spurred the development of bi-lateral bargaining solutions and market creation.[1] Standard economic analysis points here to the decisive role played by transaction and negotiation costs in explaining the patterns of bi-lateral bargaining solutions to joint and global environmental problems. This standard approach to dispute settlement between parties joining international environmental conventions underlines the role of the question of national sovereignty, opportunism and power. Dispute settlement builds first of all on dispute avoidance via monitoring, reporting and inspection, then on well specified non-compliance procedures, then on consultation and negotiation, mediation, conciliation, arbitration and, finally judicial settlement in the international Court of Justice with its special environmental branch, but all these measures depending on the voluntary participation of the parties (nation states) involved.[2] This way, it is possible for nations not only to avoid binding international agreements but also to engage in prolonged disputes over the actual interpretation of international agreements, if they choose so. On the other hand, we see a growing number of international agreements together with a growing understanding of the global and common nature of many of the environmental problems, that were earlier considered as purely local and exclusively national. In this way, we will regard in principle all kinds of cross-national, bi- and multilateral bargaining solutions as JI.

Implementation of cross-national/global environmental goals have developed into the following two main categories with their subcategories:

A. Different types of international joint implementation projects:

1. International aid and support programmes of funds for solution of environmental problems in poor countries (e.g. UN programmes)
2. International agreements to solve common environmental problems (e.g. Vienna/Montreal CFC treatises)

B. Different types of bi-lateral joint implementation projects:

1. State funds for solution of environmental problems in other countries (e.g. Western Europe countries investing in Eastern Europe countries, Northern Europe countries investing in Southern Europe countries)
2. State funds for remedial or compensatory action of own environmental problems in other countries (e.g. carbon sink projects)

3. State funds for a mix of B1 and B2 (e.g. dept-for-nature swaps)
4. Private funds for solution of B1 or B2 or B3 (e.g. environmentalist group financed conservation projects)

More concretely, the 1992 United Nations Framework Convention on Climate Change (FCCC), Article 4.2. opens for JI projects if they meet the prerequisites of contributing to the reduction of global greenhouse gas emissions, if this effect is controllable and verifiable, if the costs of achieving the emission targets are lower than purely national investments and if both the investing and the receiving countries are better off implementing the project than not, including the total costs and benefits of the project.[3]

It is within this framework possible to distinguish between four different types of projects and direct strategies towards global climate change: 1. Fossil fuel saving by substitution of energy sources. 2. Improvements of energy efficiency, change of industrial technologies and substitute for CFC's, etc. 3. Carbon sink enhancement by changes in land use and reforestation or conservation. 4. Change of agricultural practices leading to reduced methane emissions.

On a global level the projects have the potential advantages of increasing the incentives to reduce greenhouse gas emissions, develop new technologies, encourage cross-country commitments and reduce the overall costs of implementation of international targets. The donor countries can have the advantage of cost savings, getting a national share of global climate benefits and new possible investment and export markets. For the receiving countries, the advantage should be in the access to additional financial resources, transfer of technologies and possible cost savings due to new technology, getting a national share of the global climate benefits, getting national and local environmental benefits, job creation and capacity building.[4]

Some possible disadvantages of JI have also been recognized. The whole question of monitoring, control and verification of the investment projects is very complex, along with the uncertain effects on technological change and abatement efforts in the donor country and the possible distortion of the development preferences and opportunities of the receiver country, the increased foreign influence over management of natural resources and the overall global equity effects of the projects. To this should be added the possible medium and long term lock- in effects of investment projects, which in the short term may seem appropriate. The modernization of the Eastern European energy systems by means of technology transfer from e.g. Denmark is here a good example. The Danish technology may help increase the energy efficiency of existing utilities and thereby work positively towards the present CO_2 emission targets, but in a longer perspective, these investments can prove to be an obstacle for achieving tighter emission reduction targets, which would imply e.g. a substitution of fossil fuels with renewable energy sources.

The Central American forestry case

Emissions of greenhouse gases, of which CO_2 is the most important, have been growing steadily during the last 2 centuries, especially in the second half of this century. The 1990 world total emission estimated was 21.6 billion tons of CO_2 and the OECD countries had a share of approximately 50% of this.[5] In the same period we have witnessed a steady decrease in areas covered with natural forests and generally a degradation of the quality of vegetated land.

Human-induced land degradation has been estimated to amount to 244 million hectares in South America in the period 1945-1992, in North and Central America 158 million hectares, in Asia 746 million hectares, in Africa 494 million hectares and in Oceania 103 million hectares. These figures are equal with 14, 8, 20, 22 and 13 percent of total vegetated land in the regions respectively.[6] Deforestation in Central America, for instance, has in the period 1980-1989 annually been at a rate of 3.1% (Costa Rica), 0.9% (Panama), 2.3% (Honduras), 3.2% (El Salvador), 2.7% (Nicaragua), 2.0% (Guatemala) and 0.6% (Belize) in many cases for a large part due to the continued movement of the agricultural frontier together with fire wood and charcoal extraction.[7] In the 1990s the Central American deforestation amounted to 416.000 hectares per year approximately, meaning 48 hectares deforested per hour.

Taking into account these two problems: carbon emissions and deforestation, there are very good reasons behind the wish to integrate strategies towards these problems. JI refers here to reduction, fixation or avoidance of carbon emissions globally, precisely through forest conservation, management and reforestation programs. This is a new opportunity to gain additional benefits in the development of projects that reduce emissions of greenhouse gases. It refers to international arrangements through which an entity of one country partially meets its commitment to reduce greenhouse gas levels by undertaking cost effective mitigating projects in another country. This is linked to the FCCC and in the case of Central America there are already some projects being developed in Guatemala, Costa Rica and Belize with counterparts from the United States and the Netherlands among others, which are still under negotiation. The main objectives of the joint implementation programme are: 1. Identify and initiate cost-effective opportunities for reducing the rate of the global build-up of greenhouse gases. 2. Support and encourage sustainable economic and human development. By influencing the distribution of private capital, JI can steer investment projects into the path of sustainable development. 3. Promote other social and environmental objectives at the local, national and regional levels, such as the protection of watersheds, increased food production, biodiversity protection, or heightened forest yields.

Two specific types of joint implementation projects have been developed in Costa Rica: one is called 'clean' energy and the other one is forestry related projects. In the first one the idea is that a power company from another nation will produce energy through a non-emitting plant in Costa Rica, like the United States Northeast Utilities Co. has already done with a 20 megawatt wind farm there.[8] The second possibility is to invest (no to buy) in a piece of forested land, so that rather than being converted to non-forested land, it would be maintained in its natural state as capturer (sequesterer) of carbon. Another possibility related to this second possibility is investing in reforestation and another one is the correct sustainable management of a natural forested area. All these possibilities allow forest owners, which many times are poor peasants and small landowners, who have been expropriated but not paid because of fiscal deficits, to receive a partial compensation of the forest services they are providing to the country and the whole globe. These landholders receive the funds from this carbon sequestration and use them for reforestation, forest protection or sustainable forest management.

The Costa Rican carbon bonds

In a formal agreement with Norway, the Costa Rican government very recently issued carbon bonds (Greenhouse Gas Emissions Mitigation Certificates) of a value of 2 million $ whereby Norway have bought a sequestration service of 200,000 tons of carbon from Costa Rican forests. The sequestration service will be provided over a period of 25 years through reforestation and forest conservation projects in an area covering nearly 142 square kilometres in Costa Rica. The agreement between the Costa Rican and Norwegian governments are designated as a pilot project of the JI programme of the FCCC and it is estimated, that Costa Rica have approximately 400,000 hectares of degraded land, which could be reforested in a similar way.[9]

The agreement has been accompanied by a new institutional set-up in Costa Rica, where a special national office for JI, named Costa Rican Office on Joint Implementation (OCIC: Oficina Costarricense de Implementacion Conjunta) is in charge of the international negotiations and agreements. The carbon funds resulting from the sales of carbon bonds are transferred to the national forest fund (FONAFIFO: Fondo Nacional de Financiamiento Forestal) which invest in national parks, forest conservation and reforestation projects. The monitoring and control of the projects are in the hands of the National System of Conservation Areas (SINAC) in the Ministry of Environment and Energy (MINAE) together with private sector auditors. The individual landowners can then submit an application to the national forest fund for financial support to reforestation or forestry protection and FONAFIFO can, further more, make use of the funds to ensure existing national parks as well, and to compensate landowners who must meet regulations on the use of their properties. The Norwegian bonds give a value of 10$ per ton of carbon. This carbon price has been calculated on the basis of an estimated average income loss per hectare of 50$ a year in agriculture and an estimated annual carbon fixation capacity of wood land of 5 tons per hectare.

The Costa Rican carbon agreement is based on a mix of public and private Norwegian funds, where 300,000$ comes from a private consortium engaged in hydroelectric projects in the area. The Norwegian government is also engaged in a wider JI project between the two countries which includes direct Norwegian investments in a modernization of an existing hydroelectric plant, which also has mitigating effects on greenhouse gas emissions in the region. The Costa Rican forestry project copes in this way with climate change both via carbon sequestration and via watershed maintenance for hydroelectric energy production. A joint-joint and really mixed type of bi-lateral project (a new type B5, c.f. the first section) addressing environmental problems in both the donor and the receiver country with both public and private funds.

JI programmes, such as in the case of Costa Rica, also allows the generation of several other activities from the same forest without affecting the carbon storage. These examples are ecotourism, extraction of minor forest products, such as latex, fruits, wildlife, nuts, etc., the use and research of biodiversity. Additionally, each one of these activities may generate multiple income streams, because in order to internationally sell the service of carbon sequestration, there is the need of many activities around this service; for instance cartographers, Geographical Information Systems (GIS) analysts, insurance companies, foresters, engineers, economists, financial system, and other specialists. An entire new economic cluster of activities is being created around an emerging new commodity, the carbon service, which is only starting to be traded internationally.

Problems and opportunities

JI projects should be viewed from many angles and considered with respect to the questions of cost-effectiveness, environmental effects, equity and linkage dynamics and learning effects of the specific projects. Forestry projects are in this regard an especially complicated issue, in that the forest provides a whole range of services and products and, accordingly involves a wide range of actors and stakeholders. A systemic and dynamic understanding of the forest system is therefore necessary in order to avoid a strict conservationist bias in the carbon sequestration projects of joint implementation. We are in some way talking of a new rationale for the forest sector, which not only includes traditional wood products, but also many services such as the ones in the following table. Table 1 gives an indication of the many dimensions of the sector.

Table 1.
Dominant and emerging joint implementation beneficiaries of goods and services from the forest.[10]

Goods and services from forests	Benefits for the forest owner	Benefits for the country	Global benefits
0. Alternative (competing) land use	(X)		
1. Timber and wood products	X		
2. Non-wood products	X		
3. Maintenance of hydrologic cycle	Y	X	X
4. Soil and water quality conservation		X	
5. Wind and noise control		X	
6. Natural scenery		X	X
7. Recreation and ecotourism		X	
8. Cultural and religious services		X	
9. Microclimate regulation		X	
10. Macroclimate regulation	Y	X	X
11. Maintenance of ecosystems and biodiversity	Y	X	X

X: Dominant; Y: Emerging; (X): Optional.

The JI of carbon sequestration implies a transfer of funds from the beneficiaries to the land owner (Table 1, row 10) and hence local incentives for global services. The same is the case for other forest services, which could develop into a benefit not only for the owner, but also for local communities. The case of biodiversity and genetic pool maintenance is a good case, which should be taken much more seriously, not only as a bank for pharmaceutical research and as a huge unknown library, but also from the perspective of future global food supply. The Andean countries are currently experiencing large-scale erosion of local varieties of indigenous crops and wild crop relatives, in Malaysia, Philippines and Thailand, local rice, maize and fruit varieties are being replaced, in

China, of the approximately 10,000 wheat varieties in use in 1949, only 1,000 remained in the 1970s and wild groundnut and wild rice are lost, United States have lost 95% of cabbage, 91% of the field maize, 94% of the pea and 81% of the tomato varieties cultivated in the last century.[11]

In the biodiversity area (row 11), there are many new possibilities for JI to the possible benefit of also local owners and communities as well as the national economies of the countries in question. The carbon bonds agreement between Costa Rica and Norway points also to new possibilities by mixing different services in the projects, as it is here the case with macroclimate service and hydrological cycle maintenance (row 10 and row 3).

It is perhaps more interesting to consider the level in-between the forest owner and the host nation and to break down the global level into nations and communities as well. We will return below and carry this analysis a little further. At this point, we want to contrast this multiple forest use perspective with the problems raised in connection with the pure carbon sink JI strategy.

It is off cause no guarantee of lower global emissions if the involved countries have no emission targets in their energy policies and here the contribution of the everywhere growing transport sector is a joker in the game. As the FCCC establishes the 1990 emission level as the baseline for action, it seems also fair to count in the already stored carbon in existing forests. Deforestation will, anyway lead to a release of the carbon. Both Norway and Costa Rica has signed the convention and seeks to stabilize their emissions at the 1990 level by the year 2000.

Another objection to JI of greenhouse gas emissions addresses whether it is a cheap way out for industrialized countries and a new form of eco-colonialism, where industrialized countries buy the right to pollute and influence developing country policies. You can say that this is exactly the whole idea. The most cost effective means should be chosen and the receiving countries are voluntarily joining the arrangement; however, the debt-for-nature swaps may be a more appropriate target for this criticism.

Questions concerning the future effects of more widespread JI practices expose on the one hand a fear that joint implementation investments over time will replace development aid funds from the rich countries and, on the other, if there will be negative consequences for future international regimes of specific emission targets in the form of relatively higher future abatement costs if emission credits are given to the investing countries.

A set of special problems are connected with the practical implementation of carbon sinks and reforestation. These may be in relation with operationalization and cost-benefit monetarization of national economic and environmental effects, determination of proper time horizon and discount rate, project uncertainty and complex risk distribution, major problems and high costs of control and verification relative to national targets and baselines, choice of proper time horizon of greenhouse gas offset credits, etc. It seems that in the Costa Rican carbon bonds case, negotiators have dealt with these problems. Prices and time horizons have been determined and a system of monitoring and control have been organized to the apparent satisfaction of Norway. The bonds have not any international status as emission credits, which may demand some new institutions (greenhouse gas emission bank) on the arena to propel.

Cost effectiveness is the main argument behind the engagement of heavy emitting countries' engagement in joint implementation, but as a project performance criteria it is a self-evident outcome of the valuation method applied and the Norwegian case is a good example. The Costa Rican carbon price was fixed according to income loss in relation with alternative use (agriculture) and, off cause this loss is lower in the poor countries than in Norway itself – or in Denmark, for that sake. In Belize, Bolivia and Guatemala they are at present negotiating a carbon price of 0.1$ per ton and of course, these countries are even poorer. The Costa Rican agreement is, as explained, a mix of carbon fixation and watershed protection in order to secure supplies for a hydroelectric energy project, but does this explain the huge price difference? Did Costa Rica make a good deal of selling 2 services from the same forest or did they in fact give the hydrology service for free?

It seems important to address the causes of the environmental problem more directly, i.e. the greenhouse gas emissions and deforestation, respectively. In the last case it is a question of the pressure of the agricultural frontier and colonization of the forest areas, often promoted by national agricultural policies, a dominant agro-culture, weak market opportunities for alternative forest products and weak negotiating position of forestry interests.[12] The carbon bonds work as a hidden carbon tax in the donor country, but maybe it should be much more targeted to deal with the underlying causes in the first place and in this case be earmarked to developing alternative forest products as agroforestry, ecoturism etc.

The general policy lessons from Central American forestry projects points to the need of development programmes and projects to be compatible with national development interests, to be efficient in the use of national and external funds and the programmes and projects should have a greater focus on developing local capacity and participation.[13] In the area of biodiversity projects, there is called for stronger involvement of local people, development of local economic incentives as e.g. improved resource management, exploitation of marketable biological products, investing in community social services and infrastructure and nature tourism development together with a stronger enhancement of learning processes.[14]

It is not easy or in any way clear cut how the promotion of technical change and innovation is best practiced, but market signals on the demand side normally have to be supplemented by enhancement of the supply of alternative means and options. A change in view away from project-specific technology transfer (e.g. GIS, management practices, etc.) and towards technology diffusion, linkages, learning and capacity building between projects and between the forest sector and other sectors of society is important. In the same way, the development of alternative sources of income for the resource-based communities involves closer interaction and collaboration with other sectors of society, that be the scientific community, the traditional tourist sector, the manufacturers and exporters of traditional agricultural and forest products, etc.

Local and global trade offs in policy and science

The specific demands of the FCCC emphasize that the investment project must be costs efficient and both the investing and the receiving countries must be better off implementing the project than not, including the total costs and benefits of the project, which is the traditional welfare economic Pareto optimum based on cost-benefit considerations. This is possible at any level of income distribution, why the purely economic welfare criteria always are accompanied by ethical reservations and considerations about equality, fairness and justice.

To this we have added the question of development dynamics, linkages and learning. Here the experimentation with JI could open up for development, but this view challenges the green organizations. A survey among the major environmental organizations of the United States (Nature resources Defence Fund, World Wildlife Fund, Greenpeace US, Sierra Club, Friends of the Earth US) reveals an almost uniform rejection of the JI strategy towards reducing the emission of greenhouse gasses, because the JI projects must meet the criteria, that investing and receiving countries must both have quantifiable emission targets in accordance with the FCCC (stabilization at the 1990 emission level by 2000), that the industrialized countries must first have met the convention targets of the first phase running until the year 2000 themselves and that greenhouse gas offset credits can only be recognized on the basis of creditable control measurements both when the project is ended and while it is running.

The only major organization with a positive attitude towards JI is the Environmental Defence Fund, which supports the strategy because greenhouse gas problems are global, the strategy will lead to cost-effective solutions for the industrialized countries to meet the climate convention targets and, hence, minimize the economic, social and political consequences, it will support the sustainable development of the less industrialized countries by securing funds and/or technologies for environmental improvements, the strategy promotes a broader engagement in the fulfillment of the FCCC targets and it promotes technology transfer and innovation, which will reduce the costs of future demands concerning emission reductions. Furthermore, the Environmental Defence Fund views JI projects as primarily privately funded projects, so the organization assumes that these projects will be added to the publicly funded development aid and national contributions to the fulfillment of the emission targets.[15]

FAO and other international organizations have long pointed to poverty, migration and weak local communities as the underlying causes of deforestation and hence solutions should be found in addressing these problems. Here government regulation of local resources are often just causing more problems of compliance and enforcement. The respect for social and political consequences, public awareness and acceptance is as much a worthy and a necessary consideration to take into account in the poor countries as it is in the rich countries if partnership and development aid should be the result and not conservationist indulgence for business-as-usual.

"One should solve one's own contribution to global problems at the source, the cause instead of mitigation and remedial action (elsewhere) and use the funds available to improve the state of ones own environment". This popular view does not imply any disinterest in the problems of other and poor countries, which the examples of the Danish Nepenthes rainforest group of environmentalists and the Italian voluntary ecological guard federation, FEDERGEV indicates. They are both based on the collection of private donations in their home countries for conserving and management purposes in Costa Rica. It should maybe be noted, that the Nepenthes rainforest campaign has been followed up with a campaign for reforestation and natural forests in Denmark, while the Costa Rican activities have been supplemented with public Danish development aid money for buffer zone and agroforestry projects. The Italian projects are based on the idea, that annual private donations will support the public Costa Rican national park system and secure an annual flow of funds for maintenance, control and enforcement of environmental regulations.

In forestry, the NGOs have been innovative and relatively successful in developing participatory approaches, new technologies and management practices, processes of dissemination and networking, promoting farmer organizations and function as pressure groups in the national policy context of the receiver countries. There are apparently strong arguments behind a much closer collaboration between government funded development projects and NGO projects, also with a view to overcoming the deficiencies of NGO activities.[16]

Cost minimization is not valid as an overall or absolute standard for performance criteria guiding decisions on project implementation. The principle of cost-effectiveness is very static, but it is none the less impossible to reach any optimal and final result to guide decision-making. From a rich and donor-country point of view, it has also been applied both to suggest, that environmental degradation should be confined more to the poorer countries from overall efficiency reasons and, with JI, that environmental protection and restoration should be confined more to the poorer countries. Though the cost-effectiveness principle plays a strong role in the FCCC, it should be – and in fact is – subordinated other types of performance criteria. The driving force behind bi-lateral JI is a belief in partnerships of bi-lateral advantage in the absence of a global authority. The challenge is to integrate bi-lateral partnerships with a concern for the overall common advantage in the JI projects, while taking into account that mutual advantage may newer be the same as equal advantage between the parties.

On a general level, there is no clear-cut answer to the desirable future development of JI strategies. It should be found in the trade off between the individual science disciplines, which becomes necessary when the realm of certain knowledge, low decision stakes and low systems uncertainty are exceeded. In other words, we are leaving the specialized knowledge of the applied sciences and entering the field of post-normal scientific problem-solving strategies which include the use of the local knowledge of extended peer communities in the research and decision making processes.[17] JI is all around at the globally open political frontier and different projects evolve according to the needs and perception of those involved.

While the principle of mutual advantage serves as a pragmatic description of the present condition for implementing joint projects in the absence of strong international system of regulation and enforcement of common agreements, the call for new partnerships between the sciences and between the scientific community and the local resource-based community in question together with NGOs and governments also requires that new avenues of mutual advantage will be established for all the parties.

Concluding remarks

In the forestry case, an integrative and inter-sectoral perspective on the development of future joint implementation projects has to be recognized to secure locally sustainable (medium and long term) forestry projects. Here the local communities could be included as beneficiaries of forestry projects (cf. Table 1) without necessarily implying a (many places long passed) system of common property to the resource, if they are acknowledged the entitlement as stakeholder and actor in the sustainable management of the resource. The compensatory principle, which serves as the basis for implementing the PPP[18] must in this way guide the allocation of funds for remedial and mitigatory action towards global environmental problems The benefits of the forest services can not be allocated to the forest owner alone, as in the present Costa Rican pilot case with many small land owners, which must be compared with several of the other Central American countries, where the distribution of land (and forests) are looking quite different. Funds are here needed to the communities at the agricultural frontier and close to the natural forests, so that new development paths can be pursued.

Social embeddedness is understood as a nestedness of rules within other rules[19] both along a vertical and horizontal dimension, which makes sense in relation with the functioning of vertically and horizontally based forestry policies. The vertical dimension describes the relation of the individual and the household or firm, the organization and the community, communities and nation states and nations and the international political and institutional sphere. The horizontal dimension points to the embeddedness of rules within culturally shared appropriability rules of kinship, religion, affinity and broader sets of values. A way forward could be to promote the horizontal interaction and JI as an alternative to the hierarchical international projects. This is also very much an inter-sectoral perspective including actors and policies related to e.g. the agricultural and energy sectors. Though equal advantage between donor and receiver of JI projects is non-attainable, some coordination is needed to promote equal advantage among potential receivers. As a complement to the vertical principle of cost effectiveness and mutual advantage, more emphasis is needed on a horizontal principle of multiple advantage, where the local-local trade off's between different alternative forest uses very much is a local matter to undertake, as the case must be with the local communities of the the donor country affected by the decisions to employ a JI strategy towards CO_2 emissions on the account of other measures.

Notes

1 Zylicz, T.: "The Role of Economic Incentives in International Allocation of Abatement Effort", in: R. Costanza (ed.): *Ecological Economics. The Science and Management of Sustainability*, New York: Columbia University Press 1991.
2 OECD: *Approaches to Dispute Settlement in Environmental Conventions and Other Legal Instruments*, OECD Working Papers No. 95, Paris, 1995.
3 Torvanger, A.: "Prerequisites for Joint Implementation Projects under the UN Frame-

work Convention on Climate Change", *Policy Note*, 1993.3, CICERO, Oslo University.
4 Selrod, R., Ringius, L. and Torvanger, A.: "Joint Implementation-a promising mechanism for all countries?", *Policy Note*, 1995:1, CICERO, Oslo University.
5 IEA: *World Energy Outlook*. International Energy Agency, Paris, 1994.
6 CIFOR: *CIFOR's Strategy for Collaborative Forestry Research*, Center for International Forestry Research, Bogor, Indonesia, 1996.
7 UNDP: *Human Development Report 1996*, New York: Oxford University Press, 1996.
8 Lovejoy, T. E.: "Lessons from a small country", *The Washington Post*, April 22, 1997, p. A-19.
9 *Tico Times*: "C.R. Sells First Carbon Bonds to Norway". February 14, 1997, p. 10.
10 With some changes from: Segura. O., Gamez, M. & Gottfried, R.: *Politicas Forestales en Centro America: Restricciones para el desarollo del sector*, CCAB-AP, San Jose, Costa Rica, 1996.
11 FAO: "FAO's First State of the World's Plant Genetic Resources: Erosion of Biodiversity and Loss of Genes Continues; Many Genebanks Threatened", *Press Release*, 1996/9, Rome.
12 Segura, O.: "Politicas Forestales en Centro America", in: Segura, O., Kaimowitz, D. & Rodriguez, J. (eds.): *Politicas Forestales en Centro America: Analisis de las restricciones para el desarollo del Sector Forestal*, IICA-Holanda, Consejo Centroamericano de Bosques y Areas Protegidas & Programa de Desarollo Sostenible en Zonas de Frontera Agricola de Centro America, San Salvador, 1997.
13 Current, D.: *Forestry for Sustainable Development. Policy Lessons from Central America and Panama*, Department of Forest Resources, University of Minnesota, No. 188, December, 1994.
14 Wells, M.P.: "Community-based forestry and biodiversity projects have promised more than they have delivered. Why is this and what can be done?", in: Sandbukt, Ø. (ed.): *Management of Tropical Forests: Towards an Integrated Perspective*, Occasional papers from SUM No. 1/95, Centre for Development and Environment, Oslo.
15 Killingland, T.: "Den Nord-Amerikanske miljobevegelses syn pa Joint Implementation som virkemiddel for a redusere utslipp av klimagasser", *Policy Note*, 1994:3, CICERO, Oslo University.
16 Farrington, J. & Bebbington, A.: "Interactions between NGOs, governments and international funding agencies in renewable natural resources management", in: Sandbukt, Ø. (ed.): *Management of Tropical Forests: Towards an Integrated Perspective*, Occasional papers from SUM No. 1/95, Centre for Development and Environment, Oslo.
17 Funtowicz, S. O. & Ravetz, J. R.: "A New Scientific Methodology for Global Environmental Issues", in: R. Costanza (ed.): *Ecological Economics. The Science and Management of Sustainability*, New York: Columbia University Press, 1991.
18 Lindegaard, K.: "Sustainable Techno-Economic Evolution and Environmental Globalization", in: Tylecote, A. & Van der Straaten, J. (eds.): *Environment, Technology and Economic Growth: The Challenge to Sustainable Development*, Hampshire: Edvard Elgar, 1997.
19 Ostrom, E.: *Governing the Commons: The Evolution of Institutions for Collective Action*, Cambridge: Cambridge University Press, 1990.

Global solutions and local understanding

Conceptual and perceptual obstacles to global ethics and international environmental law[1]

Ulli Zeitler

Introduction. Sustainability, human rights and global concerns

The quest for sustainability in any of its current interpretations is a global concern. This presupposes a minimum agreement on basic issues: a common terminology, a shared diagnosis of present conditions, an agreed on goal or desirable development and a consensus regarding certain procedural rules. By now, the attempt to give the concept of sustainability a specific material interpretation has been almost entirely abandoned. Utilizing its rhetoric strength, the main focus is on processes of change in response to perceived environmental problems. There is no general answer to what sustainability aims at, but there is a consensus that something has to happen in order to meet present challenges to mankind and other natural beings now and in the near future. Giving occasion to reflect on actual (unsustainable) practice and to initiate certain actions, the talk about sustainability and its integration in various national and international agreements has had a valuable impact on current policy. Yet, we should be careful not to overestimate its success. The apparent consensus conceals fundamental conceptual and perceptual differences which threaten the viability of actual achievements. Similar to concepts like 'freedom' and 'responsibility', 'sustainability' is a word on a high level of abstraction and with clearly positive connotations. This opens up for ideological uses and for disregarding actual conflicts. The concept seems unfit to deal with the peculiarities of actual life which constitute the meaning of the concept in the first place. Behind the apparent agreement to promote sustainable solutions there are quite different perceptions, attitudes and interpretations of what sustainability really is about. These social and cultural differences affect both aims, methods and interpretative frameworks to an extent which makes any common talk of sustainability redundant.

The New Zealand 'Resource Management Act' from 1991 is a clear example. New Zealand, being a bicultural society, has made an admirable attempt to integrate traditional Maori values into an overall Western legal framework which aims at sustainability. The result, however, is a legal document which is conceptually inconsistent and has major problems of implementation, if its integrative intentions are taken seriously. Although there seems to be conceptual agreement between the parties, this is only achieved on the conditions of a standard Western terminology and numerous examples of translatory incorrectness. As a matter of fact, the interpretation of Maori values within a basically European interpretative framework has not effectuated a real synthesis but has largely eliminated indigenous concerns. The New Zealand situation showed that communication fails if it takes place on the conceptual (and perceptual) conditions of one party only. Moreover, these experiences, which I have described and analyzed elsewhere[2], should make us suspicious of the assumed neutrality and universal validity of our current Western conceptualizations.

Another example may be illustrative. The recent years' endeavour to make sure the observance of human rights worldwide seems to be a quite legitimate and praiseworthy task. However, there are major problems of implementation, even among countries which have signed the international agreements, which should question this whole enterprise. Although human rights *in abstracto* are acknowledged, *in concreto* they form part of a particular socio-natural context which as generic condition must be assumed to have a more fundamental and more significant level of reality. The interpretation of human rights claims can vary from positions which see them as universally and invariably applicable, i.e. as superior rules to be observed before anything else, to positions which make the observance of rights dependent on those particular circumstances which ultimately determine their legitimacy (as is visible in certain Asian cultures). If rights are claims against others to abstain from and to respect certain interferences, then we need some good reasons. Rights have to be justified; they are not a natural starting point. The good reasons could come either from our knowledge of particular circumstances or from so-called anthropological constants, i.e. from general reflections on the nature of human beings (to be delimited from non-human nature). So far, the last type of reasoning has been prevalent contrary to its lack of empirical substance and justificatory clearness. The problems only increase when we turn our attention to environmental issues, in particular to the ideas of 'a right to the environment', 'eco-rights' and 'nature's rights'. These concepts have a cultural bias which makes them and the assumed problems at least unintelligible for certain people. To include these rights in international agreements, huge implementation problems must be expected to emerge.

The following remarks are intended to highlight some of these problems in particular in relation to Japan. It will be argued that the scientific and political cooperation between Europe, North America, and Japan is in part based on different perceptions, attitudes and conceptual interpretations of the nature of environmental problems, which is concealed by an apparently unanimously shared verbal commitment. Even where this commitment is free from conceptual misunderstandings, the reasoning and motives might probably diverge to an extent which questions the viability of the enterprise. This fact may explain the continual controversies and implementation failures. As a consequence, we should reconsider the meaning of the first part of the expression, widely used in environmental policy: *'Think globally, act locally'*. If there is no shared global understanding, other than 'colonialist' extensions of Western concepts, perceptions and attitudes, then we need another strategy to deal with transnational environmental issues. This affects also the forming of future legislative initiatives, international conventions and agendas.

Japan and the nature of environmental problems

In the 1960s the people in several districts of Japan suffered tremendously under the effects of industrial pollution and vehicle emissions. The impact of this local pollution on human health conditions, when finally publicly recognized, caused instant political action. The first environmental laws in Japan's history date back to the these years of the late sixties. At the same time environmental science with a significant technological outlook was born. Today, environmental science and technology is solidly represented in public and private research institutions with extensive international cooperation. Also politically, there is a broad consensus on the relevance of major environmental issues and Japan has signed most international agreements which are concerned with nature protection and pollution control.

Nevertheless, any observer of Japanese society will soon realize that people, apart from particular responses to urgent environmental challenges, typically have no 'environmental consciousness' in the Western sense and only reluctantly engage in multi-purpose environmental organizations. Furthermore, when we analyze political statements, media interests and scientific topics it becomes clear that environmental issues are defined in terms of a catalogue of problems, reflecting major global issues such as global warming and CO_2 emissions, the regional and local effects of e.g. NOx, SOx and particle emissions, noise and vibration, and, occasionally, the verbal commitment to biodiversity claims. Leaning on globally discussed topics, the involvement in environmental research and politics is exclusively concerned with what is included in the topic list. Apart from confessions due to professional interest and diplomatic manners, deep commitments to the environmental case *as such* are largely missing. The contrast is astonishing compared with the attention and, partly, enthusiasm environmental issues are met with in certain circles of countries like Germany and the USA. This contrast holds even if we abstract from the fact that the public concern for environment issues in the West is partly created by the mass media and forms part of consumerism.

Consulting the predominant picture of Japan as a traditional Eastern culture which is known for its holistic and animistic worldview as emphasized by deep Buddhist, Taoist and Shintoist traditions, this lack of serious concern for environmental matters might surprise. Yet, the explanation has nevertheless to be found in Japan's religious culture. The differences in research traditions and political practice between Europe and Japan have obviously very deep roots in quite different historical and cultural developments, differences which have survived a very massive Westernization of Japan. As is well known, Japan has been largely isolated culturally and economically for several centuries preceding the so-called Meiji revolution in 1868. In the Meiji era a massive import of Western science, technology and philosophy has taken place which produced a research environment very similar to the one found in most European and American countries. Still today, however, this modernization is largely adapted from outside, more like an useful instrument than something which has grown out of the heart and mind of the Japanese people. The individualism and indifference which can be tracked in young people especially is probably no exception. At least these adaptive attitudes are quite conform with Japanese culture and wouldn't coexist with actual social practices (at working places, negotiation tables, in families, etc.) if they just were radical challenges of traditional values. A few examples may illustrate this point.

In the case of philosophy, an academic discipline which went to Japan with European science in the last three decades of the 19th century, it is quite surprising that that there actually is no genuine Japanese philosophy to be found at major universities. To the extent philosophy is dealt with at all, it is as an externally adapted object for study which is regarded as some kind of curiosity rather than a deeply felt social and existential need. Existential questions are dealt with pragmatically, occasionally interpreted within the framework of religious culture. The whole enterprise of philosophy seems to be an integral part of Western civilization and fundamentally alien to Japanese culture.

As is the case with philosophy, so has our growing concern with 'nature' and 'the environment' in the Western world turned into an issue of political and academic interest also to the Japanese. As can be expected from basically imported concepts, they are used in a technical sense only. What is particularly missing in modern Japanese society is the daily commitment to the overall, abstract intentions of a 'respect for *nature as such*' and a '*general* concern for *the environment*'. Although particular natural phenomena may get proper attention, 'nature' and 'the environment' can hardly play the role of being an object of human affection and intellectual devotion. Moreover, the devotion to natural phenomena is highly discriminative, which makes very abstract concepts like 'nature' and 'the environment' inappropriate. The zealous and careful concern for a small potted plant squeezed into the wall or the attentive study of a single "*sakura*" (cherry blossom) coexist naturally with the disregard of unrelated community yards, the negligence of foreign forest conditions or of 'the global environment'. Inconsiderate waste-disposals and insensitive development projects are likely to overshadow their opportune concern for particular natural phenomena. To foreigners, the behaviour of most Japanese looks at least contradictory.[3] As the Japanese typically have no relation to 'nature *as such*', as an abstract entity, nor to those parts of nature which are more or less at a distance to their immediate life context, but primarily to particular natural phenomena in some way related to their daily activities, nature conservation issues, animal rights movements and philosophies of nature make little sense, which might obstruct effective, joint measures to prevent future 'environmental' and 'nature' problems.

Some peculiarities of Japanese culture and mind are nicely reflected in the linguistic conditions and the current social structure which have been shown to be relatively stable through the last century.[4] The Japanese language had originally no word for the English concept of 'nature', though '*shizen*' has in modern times been used to equate the Indo-European 'nature', but there is a great variety of expressions for particular natural phenomena. The reason for this is, according to Susumu Ono, that 'nature' wasn't conceived to be an object vis-á-vis the human being to be exploited, dominated and cultivated. Instead, man is considered to be deeply engrossed in a particular context of nature as his existential place, his space of action.[5] This focus on particular circumstances *as lived but not objectified* is a characteristic feature of Japanese language, visible in certain distinctive traits of communication. Like Finnish, with which the Japanese often compare their language, the grammatical subject is often regarded superfluous. Furthermore, the choice of the first person personal pronoun is basically dependent on whom I am talking to and on the character of the relation which is going to be established. In Japanese the most used translations of "I" or "me" are "*boku*", "*ore*", "*onore*", "*washi*", "*oira*", "*temae*", "*jibun*", "*watashi*", "*watakushi*", "*atashi*" and "*uchi*". There is no abstract referent to the acting or relating subject as such (although language guides usually try to accommodate to the structure of European languages by appointing "*watashi*" or "*watakushi*" to such a status). In fact, in most cases there is no need to identify the individual subject by using personal pronouns. What is essential is the matter at stake in the particular relationship. Asking another person what he is doing, we normally should use the "You" and "I ". However, in Japanese the conversation would rather look like this: "What doing?" – "Listening to music", and differentiate between different social circumstances, gender and forms of politeness (e.g. *Nani wo shite imasu ka. – Ongaku o kiitemasu.* Or: *Nani shiteruno. – Ongaku o kiiterundayo.* Etc.) In these sentences the "I" and "You" are in the background, while the fact of "doing" and "listening" without subject are in the foreground. "I" and "You" do not exist as pre-given subjects which establish some relation between them. First there is the *"Between"* man and man, linguistically expressed in the word for man or human being, "*ningen*", constituted of the words "*nin*" (person) and "*ken*" or "*aida*" (between). This "Between" is neither the result of some "relating" nor is it "not yet" dissolved. And it is certainly not a Between "Man" and "Man". The togetherness and unity of "I" and "You" is a self-sufficient reality.[6]

An analysis of Japanese language so far discloses two important things: firstly, the dependence of the choice of appropriate terms and concepts on the character of the particular situation in a way which presupposes an intuitive understanding of and familiarity with its circumstances. The lack of abstract concepts in original Japanese and the peculiar linguistic inflexions are clear indicators of this trait. Secondly, and closely connected, there is the basically non-dualistic, but far from tensionless attitude, imbueing the linguistic understanding. Seeing man not as an autonomous individual but as living out of the "Between" human beings is an example of a language structure which moves beyond the subject-object dichotomy.

Having these general characteristics of the conceptual framework of the Japanese in mind, the concept of "environment", although well-established in contemporary Japanese society, must look rather bizarre. "Environment", "*kankyô*", means literally "circle line", "surrounding" ("*kan*") "borderline" ("*kyô*"), "what you are bordering on", or "circumstances". In current literature it has a very technical meaning, from its geometrical origin to the technical understanding of what Europeans have identified as global issues of ecology and nature preservation. To give the concept an ontological foundation, as 'the environment' has in current ethical and metaphysical research, cannot be justified from conceptual analysis.

Tetsuro Watsuji (1889-1960) has often been consulted to make sense of the peculiar features of Japanese attitudes to nature.[7] His famous, but widely contested climate theory is repeatedly referred to. In Bin Kimura's account, "only under the presupposition of the trustful-intrusive feeling of confidence, that nature will not punish you, you will be able to behave freely and unforced in the Japanese climate."[8] While the Europeans find their trust in nature in the rationality and lawfulness of nature, this rational guarantee is missing in the Japanese mind. Being free for natural catastrophes is rather accidental. But to live with this insecurity would be nerve-racking. That's why the Japanese need moments, where they can breathe; using "amae", i.e. presupposing the benevolence of nature and behaving as if there were no danger.[9]

This is also reflected in their attitude towards technology. Technology is for the Japanese a natural, instrumental part of their life. Their daily life is natural in a way that they wouldn't acknowledge that "nature" might be unable to deal with waste and pollution due to human (technological) activities. Living on an island, they think the garbage may blow away. Generally, neither Shintoism nor Buddhism seem to make any active contribution to current belief systems, although they are both, Buddhism more than Shintoism, compatible with the pluralism and pragmatism of modern Japan. In this society anything fits: technology, different religions, different lifestyles, as long as they enrich their lives.[10]

The earlier described subject-free nature of Japanese language is also reflected in a corresponding non-individualistic morality.[11] Generally speaking, we do not presuppose an identity on the individual level, but some kind of "consanguineous-historically constituted identity".[12] This is an identity which relates to something prior to the individual subject and from which this subject evolves: the *"Between"* man and man, or, nature in its singularity's relatedness to man.

Now, to elaborate on this focal concept, the notions of *"giri"* and *"ninjô"* are of central importance. Generally, the "Between" man and man is the ground for the obligatory force of *"giri"*. *"Giri"* means sense of duty or gratitude, but a sense of duty or gratitude which, due to its collective source (the immanent "Between" versus the European transcendent God), is not individualized. Both "actor" and "victim" are fundamentally affected in a way which not only prevents one-sided responsibilities, but also transcends causal relationships. *"Between one's guilt and the other's misfortune is no room for causal relations. Both are indistinguishable dimensions of the same event. Neither do I hurt someone because I am bad, nor am I bad because I hurt someone. To be bad and to hurt someone is one and the same thing."*[13] If I, due to some problematic action, have lost face, this is not primarily in relation to a particular person, but to the "Something Between" man and man which I owe my existence. People who have no particular *"giri"* to certain people or things may act rather cool and irresponsible towards them, while they otherwise behave correctly and conscientious.

The other central concept is *"ninjô"*, the peculiar feelings (*"jô"*) of man ("*nin*"), that is, the potential for apprehension of "*mono no aware*", a deep sympathy for things. This sympathy has two aspects: an emphatic understanding of the deepest nature of things and events, and based on this understanding, a corresponding sensation. The emphatic sympathy with people and things is beyond the sphere of good and evil, more original than the horizon of morality, a world prior to the separation of good and beautiful. The source of this sympathy is not a divine power but the "Between" man and man, the nature in its singularity's relatedness to man, or the "climate" as the place for the constitution of the person.[14]

As a consequence of the foregoing remarks, verbal communication plays actually only a minor role in human and transhuman relations. Moreover, a central feature of verbal communication is its vagueness, subtlety and intricacy.[15] The implications for situations of negotiation and decision-making are well-known and a favourite subject within management literature,[16] Japan-studies and comparative linguistic analysis.

To bring to a conclusion the reflections so far made on the conceptualizations, perceptions and attitudes of Japanese people on environmental issues, the following observations attract particular attention. At least it is generally expected that human and other natural beings interact with each other on the basis of trust and emphatic understanding. Together with a pre-moral attitude (which nevertheless implies what Europeans would term certain moral commitments), first of all a commitment to take part in the collective responsibility for what happens in one's radius of action, trust and emphatic understanding are conditioned on a well-organized social framework. Yet, our ultimate responsibility for the state of things has important limitations, because in the end things are really not under our rational control. To a certain extent, technology has given us some limited power, but technology itself is at the mercy of the unpredictable power of nature. This favours a pragmatic attitude. The import and development of Western technology is part of a pragmatic lifestyle. To try to justify this attitude, referring to philosophical assumptions and theories, might satisfy the European mind but is basically alien to the Japanese.

With this in mind, any appeal to 'human rights', 'individual causal responsibility' or 'natural duties' must probably fail. Nor is it easy to discuss issues dealing with 'sustainability'. As long as agreements can be based on particular reduction measures (e.g. CO_2, SOx, etc.), the term has a sharable technical meaning. But as a global demand on justice and human rights, the term has virtually no meaning.[17] On the other hand, from the point of view of the European mind, what makes sense are not a number of particular actions but their reference to the general idea of distributive, 'intra- and intergeneral justice'. Even short-term political interests have this appeal to an overall moral idea which is supposed to have universal application (in accordance with the rationale of European enlightenment). On a pragmatic level some kind of coordination and cooperation between Japan and Europe is quite possible. But there seems to be a major disagreement about the reasonable terms of this cooperation. It is important, therefore, to look for alternative ways to improve local, regional and global environmental conditions beyond international agreements and strategies.

Some suggestions for how to conceive and handle environmental problems – locally, regionally and globally

To proceed, a distinction between material and procedural questions, although they penetrate each other, might be fruitful in order to evade the situation of incompatible and irreconcilable material claims. Material claims will easily get into trouble if they are insensitive to the particular circumstances. For example, one could imagine that the idea of environmental utilization space is made the basis of a global distributional policy. A widely egalitarian society which is historically committed to a universalistic morality and believes in social control based on scientific progress would not only accept the suggested policy but also demand its global implementation, irrespective of other cultural heritages. This colonization policy, which is based on a principle of global justice, however, can hardly be expected to be adopted voluntarily outside the European-North American cultural context.

Yet, there might be material principles which respect these regional differences. So, even the concept of environmental utilization space could be applied under observance of specific local differences, e.g. recognizing that a particular underdeveloped region should have access to a bigger share of global resources than what it on average would be entitled to. Still, there are two important problems: one has to do with the idea of a maximum of consumption, the other with the relation between procedural and material demands. To establish certain legally enforceable maximum values is clearly unsatisfactory if the aim is to get ethically defensible or ethically right criteria. It encourages 'maximum pollution' instead of situational adequate behaviour. An example from traffic management might illustrate the point.[18] Legally enforceable speed limits in itself give road users no moral incentive to minimize their speed. On the contrary, the speed limits are exploited consistently and one's actions are considered 'right' as long as they do not perceptibly transgress the limits. It is, furthermore, recommended that we drive as close to the speed limits as possible in order to maintain a constant traffic flow and avoid hazardous overhauling. The need for situative moral judgement is suspended and the sphere of responsibility restricted. Now, if we substitute traffic speed with resource consumption (or just generalize from transport consumption to other forms of consumption) it becomes clear that even regionally or locally adjusted threshold values which legitimate particular consumption quantities lack the power to ensure *qualitatively* right solutions. To speak of *qualitatively* right solutions instead of ethically right solutions leaves open that the moral discourse could be replaced by pragmatic or religious solutions for societies (like the Japanese) which are at a loss about it.

My second point is this: if certain material claims have to be adjusted to particular regional or local conditions, then procedural question of how to deal with these claims become important. Probably it will turn out that coming to terms with whose interests are at stake and which priorities should be observed will be more important than the state of the global environment. Which significance global environmental issues have on our particular decisions and behaviour will be determined locally and within the context of a relatively narrow socio-natural community. To make this community function is at all times more important than our enlightened insight into trans-local and trans-regional conditions.

Procedural claims to promote a sustainable development may vary according to particular cultural traditions. There are several ways to handle local environmental challenges. In Europe, North America and Australia a democratic, discourse ethical approach is the most probable way to engage in priority questions. In Japan, differentiated social structures play probably a much more important role in communication. Group solidarity and collective responsibility oust the autonomous, rational agent of the West. While we should not discourage intercultural communication, a decentralization of decision processes is highly recommendable. It still might be true that sustainability only gives sense on a global scale. This does not imply, however, that actions towards sustainability have to take the form of joint global action. On the contrary, the foregoing comparative analysis has indicated that there are paradigmatic differences between cultures and traditions as to the perception of "environmental problems". To try to enforce a global consensus on these issues is not only unrealistic, but also presumably inefficient. Maybe the most efficient way to deal with environmental, social and economic problems is to start from a respect for local traditions, culturally determined rationalities and perceptions. Although in different ways, they all may contribute to more sustainable solutions. The problems emerge where the various historical, social decision mechanisms due to conflicting external influences are called into question and maybe even paralyzed. Thus, the continued Westernization of Japan and other countries is very likely to blur the situation. Still, by now important paradigmatic differences prevail. The pragmatic attitudes of the Japanese may lack the systematic and abstract approach of the Europeans, but nevertheless deal properly with their local activities on the basis of good and bad experiences. The dualistic tradition of Western people, on the other hand, necessitates dualistic solutions, i.e. a high degree of theoretical modelling, to overcome the self-caused misery.

The ecological movements have propagated that we should "think globally and act locally". The analysis showed that this formula might be in need of revision. There are quite obvious limits to the potentiality of *thinking globally*. The continued Westernization, the internalization of market economy and the formation of global organizations might eventually lead us to this condition as an empirical fact. However, at least the Japanese example shows that this Westernization has some important limits, and the increase in national states and national consciousness all over the World is another sign of de-globalization which has to be taken seriously. So the new ecological formula might be *think locally and act locally*, presupposing that only on a local level we might have a chance of truly being integrated with nature as the precondition of a sustainable development.

On the other hand, our environmental condition is so serious, globally and locally, that this local approach might be rather dangerous. Aggregate impact is difficult to detect, at least if we rely mainly on sensuous attention (non-reflective rationality). Environmental problems are trans-boundary problems and our responsibility may not be mainly an individual liability but collective in nature.[19] The identification with our sphere of responsibility, although still locally delimited, must not exclude the openness for what is going on beyond the place or sphere within which we reasonably are supposed to act. To some extent and especially in critical situations, the need for instant action, globally, demands joint agreements and joint actions cross the national and cultural borders. So, it is an important feature of international environmental law to think *"in terms of joint responsibility, regional cooperation or global common concern criteria."*[20]

Nevertheless, looking for *enduring*, sustainable solutions and for a reasonable balance between human and non-human concerns, the important thing is not the attainment of some agreed on ideal 'state of nature', but the way we *continuously* and *responsibly* deal with daily challenges in a particular socio-natural context. It is important to observe that this formulation does not already presuppose a particular culturally determined moral understanding. What it presumes is an ontological thesis which states that the interactions of human and non-human beings constitute challenges to which we in one way or other have to respond. Our response-ability determines the range of our commitments and obligations. In this sense, ethical responses are ontologically established. Within the context of different cultural traditions, challenges have been met differently. The idea of univerzalisation of our individual maxims and the insistence on natural or human rights are framed within European tradition as proper responses to the challenges of reality. Other cultures have developed less moralistic responses. The reliance on the immediate pragmatic context and our sensuous intimacy with our existential space are alternative ways to respond to given challenges.[21] So, the demand on human beings 'to deal *continuously* and *responsibly* with daily challenges in a particular socio-natural context' is taken to be an ontological fact (as an expression of normative reality), not just an invention of a rational and moral agent (though it is he who has *conceptualized* this fact).

Notes

1 This paper is based on a two months study stay at Tokyo Institute of Technology in 1996, financed by the Danish Transport Council, a seminar with the title 'Making Sense of the Ecology-Society Interface' at Pori in Finland in 1996 and a working paper 'Attitudes towards the environment, environmental research and environmental ethics in modern Japanese society'. CeSaM 1996.
2 Cf. Ulli Zeitler, 'Obstacles to an ecologically sustainable environmental law. The New Zealand experience.' Paper presented at an internordic workshop on 'Environmental law and environmental ethics', University of Oslo, Febr. 1995. Rev. Oct. 1996; and Ulli Zeitler and Ellen Margrethe Basse, 'The ethical rationale of the concept of sustainability and rationality conflicts in environmental law'. To be published in: Peter M. Christiansen et al., eds.: *Rationality and the environment,* 1998.
3 Aiko Ogoshi, 'Die dualistische Sicht der Natur in Japan'. In: Hans Kessler (ed.): *Ökologisches Weltethos,* Wissenschaftliche Buchgesellschaft, Darmstadt 1996, p.124.
4 Chie Nakame: *Japanese Society,* Weidenfeld and Nicolson, London 1970, p. 149. Cit. in: Kurt Singer: *Mirror, Sword and Jewel. The Geometry of Japanese Life,* Kodansha, Tokyo 1973. Introduction by Richard Storry, p.13.
5 Susumu Ono: *Nihongo no nenrin [The annular ring of Japan],* Tokyo 1966, p. 12f., ref. in: Bin Kimura: *Zwischen Mensch und Mensch. Strukturen japanischer Subjektivität,* Wissenschaftliche Buchgesellschaft, Darmstadt 1995, p. 83. Cf. ibid., p. 94. See also Haruko Okano, 'Das Problem Mensch und Natur im japanischen Kontext'. In: Kessler (ed.), *op.cit.,* p.136.
6 *Ibid.,* p.101-105.
7 Tetsuro Watsuji: *Fudo. Wind und Erde,* Wissenschaftliche Buchgesellschaft, Darmstadt 1992.
8 Kimura, *op.cit.,* p. 108-109.
9 *Ibid.,* p. 109.
10 This assessment of the Japanese view on technology has been put forward to me by Prof. Yasuo Sakakibara, transport economist and Buddhist priest, Chiba University at a meeting on March 12th, 1996 at the Imperial Hotel, Tokyo.

11 The account in this paragraph on virtue is greatly influenced by Kimura's book, especially ch. 2.
12 *Ibid.*, p. 14.
13 *Ibid.*, p. 51. (My translation.)
14 Cf. *ibid.*, p. 62-63 and Watsuji, *op.cit.*
15 Cf. Takeo Doi: *The Anatomy of Self*, Kodensha, Tokyo 1986, p. 33.
16 E.g. Robert March, op.cit. and Alan Goldman: *Doing business with the Japanese*, New York State University Press, New York 1994.
17 So, by and large, my attempts to introduce ethical topics in relation to environmental and transport research for Japanese colleagues have failed.
18 The example is taken from Ulli Zeitler, 'Traffic segregation and ethics'. In: *Proceedings of the Conference Road Safety in Europe*, Birmingham 1996. VTI konferens No.7A, Part 1. 1996, p. 42.
19 Cf. Harald Hohmann: *Precautionary Legal Duties and Principles of Modern International Environmental Law*, Kluwer, London 1994, p. 312 and Kenichi Yamagami & Eijun Suketake: *Japan's Constitution and Civil Law*, Foreign Press Centre, Tokyo 1994, p. 40-41.
20 Hohmann, *op.cit.*, p. 300. Cf. p. 311.
21 Cf. *supra* and I.G. Simmons: *Interpreting Nature*, Routledge, London 1993, ch. 5.

Contributors

Peder Agger, MA in Zoology. Professor in Environmental Planning with special reference to management of Biological Ressources at Roskilde University Centre, Denmark. Former chairman of the National Nature Conservation Board. Main field of research is nature conservation specifically in agricultural landscapes.

Helle Tegner Anker, Ph.D. (Law) is assistant professor at the Centre for Social Science Research on the Environment (CeSaM), Aarhus University (DK). Her research field is environmental law and she is presently undertaking a research project on "The Legal Aspects of Biodiversity" financed by the Danish Research Councils within the interdisciplinary research programme "Man, Landscape and Biodiversity".

Finn Arler, M.A. Ph.D. in Philosophy. Has researched in and lectured on various issues related to environmental ethics and human ecology since the late 70's, mainly at Aarhus University, Denmark. Research Fellow at the Humanities Research Center 'Man and Nature', Odense University, until July 1, with a project on greenhouse effect and global justice. From July 1 Research Fellow as part of an interdisciplinary research team on Man, Landscape and Biodiversity, sponsored by the Danish Research Councils, with a project on biodiversity and ethics.

Maj-Lis Follér, Pharmacist, Ph.D. in Ethnobotany. Associate Professor in Human Ecology, University of Gothenburg, Sweden. Main research area: modern and traditional medicine and health systems. Has done fieldwork among the Shipibo-Conibo in Eastern Peru. The results of these studies are presented in the book *Environmental Changes and Human Health*, Göteborg 1990.

Tim Jensen, MA, History of religions, assistant professor at Centre for Religious Studies, University of Odense, Denmark. Continuous research in present developments within the world religions, in Denmark and elsewhere. Continuous research in the 'greening of religions', in the political dimensions of religions and in the study and teaching of religion. Is p.t. involved in a study of patterns of conversion among Danish converts to Islam. Has published and edited widely.

Randi Kaarhus, Ph.D. in Social Anthropology. Works as a researcher at the Norwegian Institute for Urban and Regional Research (NIBR). Publications include *Conceiving Environmental Problems: A comparative study of scientific knowledge constructions and policy discourses in Ecuador and Norway* (NIBR's reprint 20/1996).

Klaus Lindegaard, MA, Economics, Assistant Professor, Dept. of Development and Planning, Aalborg University (AAU), Denmark. Director of Centre for Environment and Development, Aalborg University. Engaged in the joint AAU/Universidad Nacional project on Sustainable Development Strategies for Central America (SUDESCA). Ongoing PhD study on environmental economics and policy, IKE, Aalborg University, Denmark.

Kay Milton, Senior lecturer in Social Anthropology at Queen's University, Belfast, Northern Ireland. Educated at Durham University (BA in Anthropology) and Queen's University. Belfast (PhD). Research interests: Environmentalism and environmental politics, especially in Britain and Ireland and in the global arena. Main publications: *Environmentalism and Cultural Theory* (London: Routledge 1996) and an edited volume, *Environmentalism: the view from anthropology* (London: Routledge 1993).

Bryan Norton, Ph.D. in Philosophy, University of Michigan, 1970. Currently Professor of Philosophy in the School of Public Policy, Georgia Institute of Technology, Atlanta, Georgia, USA. Main research areas: inter-generational equity, sustainability theory, biodiversity policy and valuation methods. He is author of *Why Preserve Natural Variety?* (Princeton University Press, 1987), and *Ethics on the Ark*" (Smithsonian Press 1995). He has contributed to journals in several fields, including philosophy, biology, economics, and environmental management.

John O'Neill is a Reader in Philosophy at Lancaster University. His publications include *Ecology, Policy and Politics: Human Well-Being and the Natural World* (Routledge, 1993) and *Worlds Without Content: Against Formalism* (Routledge, 1991) and a number of papers in social, political and environmental philosophy.

Contributors

Poul Pedersen, Anthropologist, Associate Professor, Department of Social Athropology, Aarhus University, Denmark. Fieldwork in South India (Tamil Nadu) and Western Himalayas (Ladakh). "Nature, religion and cultural identity: The religious environmentalist paradigm" in O. Bruun & A. Kalland (eds.) *Asian Perceptions of nature: A critical approach* (London: Curzon Press)

Darrell Addison Posey is Director of the Programme for Traditional Resource Rights of the Oxford Centre for the Environment, Ethics & Society at Mansfield, College, University of Oxford, and Professor of Ethnoecology at the Federal University of Maranhao, Brazil. He is Founding President of the International Society for Ethnobiology and Executive Director of ISE's Global Coalition for Biological and Cultural Diversity He is author of over 200 books and publications, including the United Nations Environment Programme's Global Biodiversity Assessment volume entitled "Cultural and Spiritual Values of Biodiversity"

Paul Richards, is professor of Anthropology at University College, London, England, and a member of Technology and Agrarian Change Group, Wageningen Agricultural University, Netherlands. He is e.g. author of *Coping with hunger: hazard and experiment in an African rice-farming system* (London: Hutchinson*)*.

Peter Sandøe, MA & D.Phil. (Oxon.) both in Philosophy. Employed at The Royal Veterinary and Agricultural University, Department of Animal Science and Animal Health, Denmark, and chairman of the Danish Ethical Council for Animals. Since 1990 the major part of his research has been within bioethics with particular emphasis on issues related to animals and biotechnology.

Olman Segura, MA, Economics, Professor, International Centre for Economic Policy, Universidad Nacional (UNA), Costa Rica, Director of the subprogramme in Ecological Economics, Masters Programme in Economic Policy for Central America and the Caribbean and the research theme on Environment and Development. Engaged in the joint UNA/Aalborg University project on Sustainable Development Strategies for Central America (SUDESCA). Ongoing PhD study on Central American forestry policy, IKE, Aalborg University, Denmark

Avner de-Shalit, MA & Ph.D. in Philosophy. Teaches environmental ethics and political philosophy at the Department of Political Science, Hebrew University, Jerusalem, Israel. Associate fellow at the Oxford Centre for Environment,

Ethics and Society, University of Oxford, England. Has publiced *Why Posterity Matters. Environmental policies and future generations*, (London: Routledge 1995).

Ingeborg Svennevig, MA, Ethnography and Social Anthropology. Research fellow at Humanities Research Center, 'Man and Nature', Odense, Denmark. Doing research on the combination of local involvement and international cooperation in nature conservation in the Luangwa Valley, Zambia and the Wadden Sea area, Denmark and the Netherlands.

Merete Sørensen, M.A. in Philosophy and Humanistic aspects of computer science. Free lance lecturer. Research topic is environmental ethics. Member of the Danish Ecological Council.

Ulli Zeitler, degrees from Aarhus University in Philosophy and History. Since 1984 teacher in philosophy at Aarhus and Copenhagen University. From 1994 research associate at the Centre for Social Science Research on the Environment (CeSaM)/Aarhus. Research field: transport ethics. Various articles within the fields of environmental ethics, philosophy of law, political philosophy and management theory.